Mona Maze

Modeling of wheat growth and yield under the expected climate change

Mona Maze

Modeling of wheat growth and yield under the expected climate change

Südwestdeutscher Verlag für Hochschulschriften

Impressum / Imprint
Bibliografische Information der Deutschen Nationalbibliothek: Die Deutsche Nationalbibliothek verzeichnet diese Publikation in der Deutschen Nationalbibliografie; detaillierte bibliografische Daten sind im Internet über http://dnb.d-nb.de abrufbar.
Alle in diesem Buch genannten Marken und Produktnamen unterliegen warenzeichen-, marken- oder patentrechtlichem Schutz bzw. sind Warenzeichen oder eingetragene Warenzeichen der jeweiligen Inhaber. Die Wiedergabe von Marken, Produktnamen, Gebrauchsnamen, Handelsnamen, Warenbezeichnungen u.s.w. in diesem Werk berechtigt auch ohne besondere Kennzeichnung nicht zu der Annahme, dass solche Namen im Sinne der Warenzeichen- und Markenschutzgesetzgebung als frei zu betrachten wären und daher von jedermann benutzt werden dürften.

Bibliographic information published by the Deutsche Nationalbibliothek: The Deutsche Nationalbibliothek lists this publication in the Deutsche Nationalbibliografie; detailed bibliographic data are available in the Internet at http://dnb.d-nb.de.
Any brand names and product names mentioned in this book are subject to trademark, brand or patent protection and are trademarks or registered trademarks of their respective holders. The use of brand names, product names, common names, trade names, product descriptions etc. even without a particular marking in this works is in no way to be construed to mean that such names may be regarded as unrestricted in respect of trademark and brand protection legislation and could thus be used by anyone.

Coverbild / Cover image: www.ingimage.com

Verlag / Publisher:
Südwestdeutscher Verlag für Hochschulschriften
ist ein Imprint der / is a trademark of
OmniScriptum GmbH & Co. KG
Heinrich-Böcking-Str. 6-8, 66121 Saarbrücken, Deutschland / Germany
Email: info@svh-verlag.de

Herstellung: siehe letzte Seite /
Printed at: see last page
ISBN: 978-3-8381-3747-6

Zugl. / Approved by: Freising, TUM, Diss., 2012

Copyright © 2013 OmniScriptum GmbH & Co. KG
Alle Rechte vorbehalten. / All rights reserved. Saarbrücken 2013

Acknowledgment

It would not have been possible to write this doctoral thesis without the help and support of the kind people around me, to only some of whom it is possible to give particular mention here.

Above all, thank ALLAH for giving me courage and support in order to accomplish the task of my thesis.

I am truly indebted and thankful to my great family. My parents, brother and sisters have given me their unequivocal support throughout, as always, for which my mere expression of thanks likewise does not suffice.

This dissertation would not have been possible without the help, support and patience of my principal supervisor, Prof. Dr. Urs Schmidhalter, not to mention his advice and unsurpassed knowledge of his research area. The good advice, support and friendship of my second supervisor, PD Dr. Eckart Priesack, has been invaluable on both an academic and a personal level, for which I am extremely grateful.

I would like to acknowledge the financial, academic and technical support of the Technische Universität München – Weihenstephan and "Chancengleichheit für Frauen in Forschung und Lehre" scholarship. I also thank the DAAD (Deutscher Akademischer Austausch Dienst) scholarship for the financial support of my research at the first two years.

I am obliged to all my colleagues in the chair of Plant Nutrition – Weihenstephan, who supported me a lot not only on the academic level but also on the personal one, with the deepest gratitude.

I am truly indebted and thankful to the Modelling Soil-Plant-Atmosphere Systems team, Institute of Soil Ecology, Helmholtz Zentrum München for their valuable support.

I owe sincere and earnest thankfulness to Dr. Harald Maier at the Deutscher Wetterdienst – Weihenstephan for providing me with the future weather information.

I would like to show my sincere gratitude to Prof. Mahmoud Medany, the Professor of agrometeorology, who guides me always at the beginning of my research career.

It is great pleasure to thank all my friends and colleagues who helped me write my dissertation successfully.

Last, but by no means least, I dedicate my research to my first homeland Egypt and my second homeland Germany, and I wish the thesis to be valuable and useful to the scientific research.

Table of Contents

List of Tables IV

List of Figures V

List of Maps VII

List of Photos VIII

List of abbreviations and symbols IX

1   Introduction 1

    1.1   The affected factors on wheat growth 1

    1.2   Temperature effect on winter wheat 1

    1.3   Temperature and phenological development 2

    1.4   Lethal temperatures 3

        1.4.1   Hot episodes effect on winter wheat 3

        1.4.2   Freezing effect on wheat 5

    1.5   Solar radiation and precipitation role in wheat growth 7

    1.6   Wheat growth related to the different soil attributes 8

    1.7   The expected climate change effect on wheat crop 9

    1.8   The effect of the expected temperature increase on wheat growth 9

    1.9   The direct effect of the expected increase of CO_2 10

    1.10   Wheat crop models 12

    1.11   Climate prediction 15

2   Materials and Methods 17

    2.1   Experimental data set 17

    2.2   The methodology of the wheat crop model 19

    2.3   Evapotranspiration and Soil Water Flow 19

        2.3.1   Potential Evapotranspiration 19

        2.3.2   Actual Transpiration 20

2.3.3	Soil water flow	21
2.4	Plant Growth	24
2.4.1	Phenological Development	24
2.4.2	Crop Growth	25
2.4.3	Environmental Factors	26
2.4.4	Partitioning to Crop Organs	27
2.5	Model calibration	28
2.6	Model validation	29
2.7	Model efficiency	29
2.8	Future weather data	30
2.9	The expected future geographical distribution of yield	31
3	Results	33
3.1	Model efficiency	33
3.2	The simulated biomass partitioning	41
3.3	Soil water content sensitivity analysis	49
3.4	Scenario simulation based on climate projection for 2021-2100	50
3.5	Spatial yield distribution in Bavaria	78
3.6	The computer-based model description	87
4	Discussion	95
4.1	Temperature and phenological development	95
4.1.1	Cardinal temperature for the developmental stages	95
4.1.2	Growing degree days for the developmental stages	96
4.2	Extreme temperature effect on the growth	99
4.2.1	The effect of hot episodes	99
4.2.2	Freezing effect	101
4.3	Water deficiency effect on the growth	103
4.4	The dry matter allocation at different developmental stages	107

- 4.5 Climate change effect on winter wheat ... 108
  - 4.5.1 The used weather climate models ... 109
  - 4.5.2 The predicted yield with different regional climate models ... 109
  - 4.5.3 The expected temperature increase effect on the growth ... 112
  - 4.5.4 The effect of precipitation decline on the growth ... 113
  - 4.5.5 The direct CO_2 effect ... 113
- 4.6 Future expected yield spatial distribution in Bavaria ... 116

5 Summary ... 119

6 Zusammenfassung ... 121

7 References ... 124

8 Appendix ... 146

List of Tables

Table 1 The details of the studied experimental sites and its distance to the used weather stations.

Table 2 The soil type and the planting date for the studied seasons at the studied sites.

Table 3 The base temperature (°C) and the accumulated thermal units (°Cd) for each developmental stage during the winter wheat crop season.

Table 4 The relative temperature effect on relative consumption rate of the assimilate reserves (TGTB) values at different temperatures.

Table 5 The plant organs coefficients and their values.

Table 6 The Nash-Sutcliffe efficiency (NSE) and the root mean square error (RMSE) for the Observations' Standard Deviation Ratio (RSR) values for the studied locations.

List of Figures

Figure 1 Comparison of simulated (——) and measured values (the symbols) for grain yield of different wheat cultivars grown at Würzburg (a), Donau-Ries (b), Freising (c), Passau (d), Regensburg (e), Lichtenfels (f), Eichstätt (g), Landshut (h), Neumarkt (i), Weissburg-Gunzenhausen (j), Main-Spessart (k), and Günzburg (l) during the seasons from 1999/2000 to 2008/09.

Figure 2 Comparison of simulated (■), measured (——), +20% of the measured values (- - -) and -20% of the measured values (— ·) for the grain yield average of the used cultivars vs. the measured average grain yield at Würzburg (a), Donau-Ries (b), Freising (c), Passau (d), Regensburg (e), Lichtenfels (f), Eichstätt (g), Landshut (h), Neumarkt (i), Weissburg-Gunzenhausen (j), Main-Spessart (k), and Günzburg (l) during the seasons from 2000/01 to 2008/09.

Figure 3 Simulated course of biomass partitioning to leaves (——), stem (——) as well to grain (——) and straw (——) yield at Landshut in the seasons 2000/01 (a), 2001/02 (b), 2002/03 (c), 2003/04 (d), 2004/05 (e), 2005/06 (f), 2006/07 (g), 2007/08 (h), and 2008/09 (i).

Figure 4 Simulated course of biomass partitioning to leaves (——), stem (——) as well to grain (——) and straw (——) yield at Landshut in the seasons 2000/01 (a), 2001/02 (b), 2002/03 (c), 2003/04 (d), 2004/05 (e), 2005/06 (f), 2006/07 (g), 2007/08 (h), and 2008/09 (i), by using a sandy loam soil.

Figure 5 The expected average yield at the future periods (1) 2021-2050, (2) 2051-2080, and (3) 2071-2100 for the models REMO (■), CLM ( ), and STARII(■) at Würzburg (a), Donau-Ries (b), Freising (c), Passau (d), Regensburg (e), Lichtenfels (f), Eichstätt (g), Landshut (h), Neumarkt (i), Weissburg-Gunzenhausen (j), Main-Spessart (k), and Günzburg (l).

Figure 6 The expected average grain yield at the future periods (a) 2021-2050, (b) 2051-

2080, and (c) 2071-2100 for the models REMO (■), CLM (□), and STARII(▨)
at:
1. Donau-Ries
2. Eichstätt
3. Freising
4. Günzburg
5. Landshut
6. Lichtenfels
7. Main-Spessart
8. Neumarkt
9. Passau
10. Regensburg
11. Weißenburg-Gunzenhausen
12. Würzburg

Figure 7 Comparison between the monthly average maximum temperature between the future predicted periods 2021 – 2050 (— ·), 2051 – 2080 (- - ·), and 2071 – 2100 (——) at Würzburg (a), Donau-Ries (b), Freising (c), Passau (d), Regensburg (e), Lichtenfels (f), Eichstätt (g), Landshut (h), Neumarkt (i), Weissburg-Gunzenhausen (j), Main-Spessart (k), and Günzburg (l) for CLM model.

Figure 8 Comparison between the monthly average minimum temperature between the future predicted periods 2021 – 2050 (— ·), 2051 – 2080 (- - ·), and 2071 – 2100 (——) at Würzburg (a), Donau-Ries (b), Freising (c), Passau (d), Regensburg (e), Lichtenfels (f), Eichstätt (g), Landshut (h), Neumarkt (i), Weissburg-Gunzenhausen (j), Main-Spessart (k), and Günzburg (l) for CLM model.

Figure 9 The timing of the initiation growth and death of the different plant organs during the different crop growth (Zadoks) stages in winter wheat (Rawson and Macpherson, 2000).

List of Maps

Map 1 The administrative districts of Bavaria.

Map 2 The estimated wheat yield distribution in Bavaria by using the REMO model at the first (a), second (b) and third (c) future period.

Map 3 The distribution of the estimated number of hot episodes in Bavaria for the REMO model at the first (a), second (b) and third (c) future period.

Map 4 The distribution of the estimated grain-filling duration in Bavaria for the REMO model at the first (a), second (b) and third (c) future period.

Map 5 The estimated wheat yield distribution in Bavaria by using the CLM model at the first (a), second (b) and third (c) future period.

Map 6 The distribution of the estimated number of hot episodes in Bavaria for the CLM model at the first (a), second (b) and third (c) future period.

Map 7 The distribution of the estimated grain-filling duration in Bavaria for the CLM model at the first (a), second (b) and third (c) future period.

List of Photos

Photo 1 Model form screen displaying selected input variables under current weather options.

Photo 2 Screen shot displaying daily output values of the growth of different plant organs under current weather conditions.

Photo 3 Screen shot displaying 10-days output values of the growth of different plant organs under current weather conditions.

Photo 4 Screen shot displaying the final output of the growth of different plant organs under the current weather data.

Photo 5 Yield comparison between different seasons under current weather data.

Photo 6 Screen shot displaying selected input variables under future weather options.

Photo 7 Model form screen with future weather options by selecting only one future weather model.

Photo 8 Screen shot of the expected biomass performance during the selected future period

Photo 9 Screen shot of the expected yield performance during the selected future period.

Photo 10 Screen shot displaying expected individual yields at the selected future weather period.

Photo 11 Screen shot of the expected average yield of the selected future weather period.

List of abbreviations and symbols

$1 - f_c$	The average exposed soil fraction not covered (or shaded) by vegetation
AMTMPT	Temperature dependent variable
AS	Assimilate flow
AWC	Available water content
AWI	Alfred Wegener Institute, Bremerhaven
BM	Crop biomass [g.dm m^{-2} d^{-1}]
CERA	Climate and Environmental Retrieving and Archiving
CGR_i	Crop growth rate at day i
CLM	Regional climate model CLM
CO_2	Carbon dioxide
CR	Consumption rate of assimilation [g.CH$_2$O m^{-2} d^{-1}]
CT	Cardinal temperature [°C]
D_{ei}	cumulative depth of evaporation (depletion) from the soil surface layer at the end of day i-1 (the previous day) [mm]
DKRZ	Climate Computing Centre
$DP_{e,i}$	deep percolation loss from the topsoil layer on day i [mm]
e	Environmental vapor pressure [hPa]
$e°T_{max}$	Saturated vapor pressure at the maximum temperature [kPa]
$e°T_{min}$	Saturated vapor pressure at the minimum temperature [kPa]
e_a	Actual vapour pressure [kPa]

EC	Canopy evapotranspiration
e_s	Saturation vapour pressure [kPa]
es_o	Reference saturation vapor pressure
ET_0	Potential evapotranspiration [mm day^{-1}]
ET_c	Crop evapotranspiration
f_{clay}	Fraction of the clay in the soil
f_{Corg}	Fraction of the organic carbon in the soil
f_{ew}	The fraction of soil surface from which most evaporation occurs
FLV	Partitioning fraction of the leaf
f_r	Frost effect factor
f_{sand}	Fraction of the sand in the soil
f_w	The average fraction of soil surface wetted by irrigation or precipitation
G	Soil heat flux density [MJ m^{-2} day^{-1}]
G_f	Glucose requirement factor
GCM	Global climate model
G_{DM}	Grain dry matter [g.dm m^{-2} d^{-1}]
GHGs	Greenhouse gases
GR_{SM}	Growth rate of structural dry matter [g.dm m^{-2} d^{-1}]
G_{rsp}	Growth respiration [g.CO$_2$ m^{-2} d^{-1}]
GUI	Graphical user interface
h	The pressure head [cm]

I_i	Irrigation depth for the part of the surface that is wetted [mm]
K_c	Crop coefficient
$K_{c,max}$	The upper limit of the evaporation and transpiration from any cropped surface
$k_{c,min}$	The minimum value of the crop coefficient (K_c)
K_{cb}	Basal crop coefficient
K_e	Soil water evaporation coefficient
K_r	Dimensionless evaporation reduction coefficient
K_s	Transpiration coefficient
L_{DM}	Leaf dry matter [g.dm m^{-2} d^{-1}]
LfL	Bavarian State Research Center (Bayerische Landesanstalt für Landwirtschaft)
LfU	Bavarian Land Office for the Environment (Bayerisches Landesamt für Umwelt)
lv	Latent heat of vaporization of water (2.5 * 10^6 J. kg^{-1}),
m_i	The ith simulated value for that being evaluated
MT	Daily maintenance [g.CH$_2$O m^{-2} d^{-1}]
NSE	Nash-Sutcliffe efficiency
$\bar{o}$	The mean values of the real data
o_i	The ith real value being collected
p	Average fraction of Total Available Soil Water (TAW)
P_E$_{10}$	The difference between the actual precipitation and crop evapotranspiration each 10 days

P_a Atmospheric pressure [kPa]

PAW   Plant available soil water content

P_i Daily precipitation amount [mm] on day i

PIK   Climate Impact Research Institute in Postdam

pt   Plant height [m]

PTFs   Pedotransfer functions

RCM   Regional climate model

RCR_{RES} The relative consumption rate of the assimilate reserves

R_{DM} Root dry matter [g.dm m^{-2} d^{-1}]

REMO   Regional climate model REMO

RES   Assimilate pools [g.dm m^{-2} d^{-1}]

REW   cumulative depth of evaporation (depletion)

RH   Relative humidity [%]

RMSE   Root mean square error

R_n Net radiation at the crop surface [MJ m^{-2} day^{-1}]

RO_i Precipitation run off from the soil surface [mm]

$R_{s,i}$ Water reserves in the soil

RSR   Root mean square error (RMSE) – Observations' Standard Deviation Ratio (SR)

RT   The relative temperature increase during the season [°C]

Rv   Gas constant for water vapor (461.5 J.K.kg^{-1})

RWLV   Growth rate of leaf biomass

S_D	Soil water deficits
S_{BM}	Structural biomass [g.dm m^{-2} d^{-1}]
S_{DM}	Stem dry matter [g.dm m^{-2} d^{-1}]
SR	Standard deviation ratio
STARII	Statistical climate model STARII
T	Mean daily air temperature at 2 m height [°C]
T_k	Temperature [Kelvin]
T_{avg}	Average temperature [°C]
T_{base}	Base temperature [°C]
T_d	Dew point temperature [Kelvin]
TDU	Thermal degree units
TEW	Maximum cumulative depth of evaporation (depletion)
TGTB	Temperature dependent variable
T_{Lmax}	Lethal maximum temperature [°C]
T_{Lmin}	Lethal minimum temperature [°C]
T_{max}	Maximum temperature [°C]
T_{min}	Minimum temperature [°C]
T_o	Reference temperature (273.15 Kelvin),
T_{opt}	Optimum temperature [°C]
T_{optl}	Lower optimum temperature [°C]
T_{optu}	Upper optimum temperature [°C]

U_2	Wind speed at 2 m height [m s^{-1}]
WDCC	World Data Center for Climate
WRSI	Water requirement satisfaction index
x_T	Extreme temperature effect factor
Z_e	The depth of the soil surface soil layer
α, n, m	Moisture retention characteristic parameters
γ	Psychrometric constant [kPa °C^{-1}]
Δ	Slope vapour pressure curve [kPa °C^{-1}]
θ	Average soil water content for the effective root zone at day (i-1)
$θ_1$	Sub-optimal temperature [°C]
$θ_2$	Supra-optimal temperature [°C]
$θ_{FC}$	Volumetric percentage of the soil water at field capacity
$θ_r$	Residual volumetric water content respectively [cm^3 cm^{-3}],
$θ_s$	Saturated volumetric water content respectively [cm^3 cm^{-3}],
$θ_{WP}$	Volumetric percentage of the soil water at wilting point
λ	Latent heat vaporization [MJ kg^{-1}]
$ρ_s$	Bulk density [g cm^{-3}]
$σ_0$	Standard deviation of the real data
φ	CO2 production factor

1 Introduction

Wheat (*Triticum aestivum* L.) is the single most important crop on a global scale in terms of total harvested weight and amount used for human and animal nutrition (FAO, 1996). Wheat is generally considered to enjoy an optimum temperature range of 17 – 23°C over the course of an entire growing season, with a T_{min} of 0°C and T_{max} of 37°C, beyond which growth stops (Porter and Gawith, 1999), whilst cultivars seem to differ in their tolerance to extreme temperature (Pomeroy and Fowler, 1973; Blum and Sinmena, 1994; Páldi et al., 1996). Wheat seems to have a lethal low temperature of $-17.2 \pm 1.2°C$, and a lethal high temperature of 47.5°C (Porter and Gawith, 1999). Grain yield depends on weather in the growing season, and how farmers choose to fertilize and protect their crops. Wheat simulation models are being used increasingly to assess the crop performance according to different environmental factors for monitoring the important results such as the grain yield, yield variability, and geographical distribution of the crop. Yield prediction is also important in assessing and managing the tradeoff between food security and the environmental impact of agricultural inputs (e.g. N pollution). The timing of certain plant disease epidemics relative to the growth stage of the crop is important (Lawless and Semenov, 2005).

1.1 The affected factors on wheat growth

The winter wheat crop performance is affected by the different environmental factors such as the weather factors (temperature, solar radiation, and precipitation), soil attributes, water supply, and the management development. This interaction between the crop growth and the different environmental factors shows a wide diversity of the crop growth and crop yield at different spatial and temporal variation.

1.2 Temperature effect on winter wheat

The temperature is one of the influential environmental factors on the wheat growth. Temperature sensitivity varies not only between the different wheat cultivars, but also changes between the plant components (Musich et al., 1981) and during the course of development. Thus, base and optimum temperature thresholds increase with development (Lumsden, 1980; Angus et al., 1981; Slafer and Savin, 1991; Slafer and Rawson, 1995b).

Therefore, the different plant organs response to the temperature differs from each other, where we can find that root growth is considered by Nielsen and Humphries (1996) to be more sensitive to temperature than that of above-ground plant parts, where the range between T_{min} and T_{max} for roots is smaller than for shoots and leaves. The optimal soil temperature for growth of the roots of wheat plants during the vegetative stage is below 20°C (Nielsen and Humphries, 1996; MacDowell, 1973) and temperatures higher than 35°C have been shown to reduce terminal root growth and accelerate its senescence (Wardlaw and Moncur, 1995). Root growth may cease if soil temperature drops below 2°C (Petr, 1991). Studies have shown an air temperature of – 20°C to be lethal for root survival (Drozdov et al., 1984). As for the leaves, Miglietta (1989) found that the T_{min} for leaf initiation is 2.5°C, where Cao and Moss (1989) found T_{opt} for leaf emergence ranged from 21.3°C to 24.3°C; values which concur with the T_{opt} value of 22°C from Slafer and Rawson (1995a). Temperatures higher than 25°C have been found to inhibit leaf appearance (Porter and Gawith, 1999). As well as, for the stem elongation the optimum temperature is considered to be 20°C, below this temperature the elongation will be slower during the vegetative phase, but with a T_{max} seemingly only slightly higher than T_{opt}. Porter and Gawith (1999) in their study showed that shoot growth between terminal spikelet and flowering was not modified by temperatures up to 16°C but was significantly reduced by temperatures above 19°C.

1.3   Temperature and phenological development

Wheat is less sensitive to temperature during its vegetative phase than during its reproductive phase (Entz and Fowler, 1988), but there is no phase during which temperature does not modify the development (Slafer and Rawson, 1994). The cardinal temperature ranges for the wheat crop at the different developmental stages during the crop growth are acting in a linear relationship between temperature and development from emergence to anthesis (Slafer and Rawson, 1995b), but generally, the later the phase of development or process, the higher the base temperature (Angus et al. 1981; Del Pozzo et al. 1987; Porter et al. 1987; Slafer and Savin 1991). For the period from sowing to emergence Porter and Gawith (1999) have shown that the T_{min} ranges from 2.4 to 4.6°C, T_{opt} from 20.3 to 23.6°C and T_{max} from 31.8 to 33.6°C, and Russell and Wilson (1994) add that soil temperature should be above 5°C. Slafer and Rawson (1995c) showed that T_{opt} of the double-ridge stage is 20°C, while Slafer and Savin (1991) mentioned that T_{min} is 4°C. The terminal spikelet stage had a lower optimum and minimum temperature range than the other stages, where spikelets may be initiated at temperatures higher than 1.5°C (Slafer and Rawson, 1995b), and the optimum temperatures for this phase lie between 9.3 and 11.9°C, with temperatures greater than 25°C being sub-optimal (Porter and Gawith, 1999). As for the anthesis T_{min} seems to be about 9.5°C (MacDowell, 1973; Slafer and Savin, 1991; Russell and Wilson, 1994) with T_{opt} between 18 and 24°C (Russell and Wilson, 1994). This is consistent with the pattern that temperature tolerance increases as plants develop, cardinal temperatures are generally highest during grain-filling (Porter and Gawith, 1999). T_{min} values for grain-filling range from 4.1°C (Hunt et al., 1991) to 8.9°C for spring wheat (Angus et al., 1981) and to 12°C for winter wheat (Russell and Wilson, 1994). T_{opt} reportedly lies between 19.3 and 22.1°C and T_{max} between 33.4 and 37.4°C (Porter and Gawith, 1999). Cultivar differences in temperature sensitivity during grain-filling can extend to one cultivar being 35% more temperature-sensitive than another (Marcellos and Single, 1972; Porter and Gawith, 1999).

1.4 Lethal temperatures

At temperatures close to the optimal temperature for growth and development plant resistance are usually maintained at the same level and fluctuations in resistance are due to internal factors rather than external ones. This range of optimum temperatures is referred to as the range of 'background temperatures'. The ranges of cold and heat hardening temperatures are to the right and left of the background temperature ranges: when plants are exposed to temperatures in these ranges their resistance increases. More extreme temperatures than the hardening temperatures are in the ranges of cold and heat injury respectively (Drozdov et al., 1984).

1.4.1 Hot episodes effect on winter wheat

Temperature is one of the environmental weather factors, which is central to how climate influences the growth and yield of crops. The rate of many growth and development processes of crop plants is controlled by air or soil temperature. Nevertheless, an increase in mean seasonal temperature of 2–4°C reduces the yield of annual crops of determinate growth habit, such as wheat (Wheeler et al., 1996b; Batts et al., 1997), grown in well watered conditions. Much of this decline in yield is due to shorter crop duration at these warmer temperatures, where a rise in temperature may increase the developmental rate of the crop, thus shortening the growing season, and resulting in a negative effect on crop production (Peiris et al., 1996). Moreover, the effects of variability in temperature on crops may also be important. First, the effects of weather variables on crops are often non-linear because of the ways in which many crop processes respond to the environment (Semenov and Porter, 1995). Second, fluctuations of extreme temperatures may affect the survival of crop plants or plant organs (Porter and Gaiwth, 1999). Under such extreme conditions, crop plants are more severely affected than under high temperature or thermal stress. However, the impact on crop yield cannot simply be predicted from the absolute temperature. Instead, it is reflected by the combination of the magnitude and duration of the hot temperature episode, and coincidence with the development stage of the crop (Wheeler et al., 2000). Thus, wheat is less sensitive to temperature during each vegetative phase than during its reproductive phase (Entz and Fowler, 1988), but there is no phase during which temperature does not modify development (Slafer and Rawson, 1994). Therefore, it has been found that high temperatures during early spike development reduced the number of spikelets per head or the number of seeds per spikelet (Johnson and Kanemasu, 1983), and temperatures higher than 31°C and lower than 9°C during anthesis may therefore be considered as the limits of successful anthesis (MacDowell, 1973; Russell and Wilson, 1994), while the rate of increase in grain dry weight increases with temperature. But both temperature sensitivity and growth rates vary between cultivars during grain-filling, where the timing of heat event during the grain-filling appears to be important, exerting a particular influence on grain quality via the accumulation of protein (Porter and Gawith, 1999). As well as, high temperatures would be expected to cause seed degradation in addition to lowering germination rate, and these negative impacts would be expressed to a greater degree for slower-germinating subpopulation (Roberts, 1988; Hardegree, 2006).

It is clear that changes to the variability of temperature, separate to changes in mean seasonal temperature, affect the yield of annual crops. The effects of brief episodes of hot temperatures on the number of yield components can be particularly dramatic. However, the impact of crop yield cannot simply be predicted from the absolute temperature. Instead, it is reflected by the combination of the magnitude and duration of the hot temperature episode, and coincidence with the development stage of the crop (Wheeler et al., 2000). Thus, isolated incidents of extreme hot or cold temperatures could seriously damage a plant. A continuous period of extreme hot or cold temperature could be lethal not only for crops, but also for humans. Summer 2003 was recorded as the hottest in Europe since 1500 (Poumadere et al., 2005).

Temperature responses may also be determined for plant processes, such as enzyme activities and photosynthesis, the rates and efficiencies of which are temperature-dependent. An example is the thermal kinetic window which describes the temperature range for optimal enzyme functioning (Makan et al., 1987). The thermal kinetic window for wheat has been identified by Burke et al. (1988) as lying between 17.5 and 23°C. Optimal rates of photosynthesis in wheat (cv. Crako) are, however, broader than these, being optimized at 25°C and declining at temperatures lower than 15°C and higher than 30°C (Wardlaw, 1974).

1.4.2 Freezing effect on wheat

Not only the high temperature can lead to a negative response to the crop growth, but both high and low temperatures decrease the rate of dry matter production and, at extremes, can cause production to cease (Grace, 1988). Low temperature is one of the primary stresses limiting the growth and productivity of winter cereals. To cope with low-temperature stress, winter cereals have evolved adaptive mechanisms that are temperature regulated. Vernalization response and low temperature acclimation are the most important of these winter survival mechanisms. Both are regulated through complex physical and biochemical interactions that are dependent on genotypic and environmental factors (Fowler et al., 1996). Low-temperature acclimation is a cumulative process (Andrews 1960; Roberts 1979; Gusta et al. 1982) that can be stopped, reversed, and restarted. The threshold temperature for the initiation of low-temperature acclimation is approximately 10°C (Olein 1967; Alden and Hermann 1971), and there is an inverse relationship between temperature and acclimation rate between 10 and 0°C (Paulsen 1968; Limin and Fowler 1985). Acquired low-temperature tolerance is rapidly lost when cereals are exposed to crown temperatures above 10°C (Gusa et al. 1982). The warmer the crown temperature between 10 and 18°C, the more rapidly the low-temperature tolerance is lost (Fowler et al., 1996).

Drozodov et al. (1984) reported that temperatures of − 2°C injured the leaves of unhardened winter wheat plants and − 4°C was lethal, but this threshold could be increased to − 13°C if plants were cold-hardened. The stem may also be damaged and weakened by frost, which may lead to lodging later in the season (Pittman, 1933; Banath and Single, 1976). Stem nodes have been reported to stop ice fronts passing through a plant internally (Single, 1964;

Ashworth and Abeles, 1984). However, the developing ear is protected at this stage within the leaf sheath, which may prevent external ice crystallisation damaging the ear. While wheat is reputed to be most susceptible to frost damage when the ear has emerged, as ice may form directly on the reproductive tissue when the ear is no longer protected by the presence of the stem and leaf sheath (e.g. Single and Marcellos, 1974; Single, 1984; Loss, 1987; Cromey et al., 1998). Therefore, the crop should be managed so that the more susceptible stages of crop development, such as the formation of reproductive primordia, do not coincide with a period when the risk of a damaging frost occurring is still high (Whaley, et al., 2004).

Winter wheat maintains growth at temperatures down to 2-3°C (Porter & Gawith, 1999). Elevated temperature has a positive effect on growth, increasing leaf initiation, leaf emergence, and stem elongation (Porter & Gawith, 1999) and a direct positive effect of predicted climate warming on growth is expected. Temperate C3 grasses maintain a significant photosynthetic activity at temperatures around the freezing point, and a minimum temperature for photosynthesis of - 4°C has been documented for some species (Skinner, 2007). Hence, winter wheat is expected to maintain a positive net assimilation even under low irradiance at air temperatures just below 0°C. Winter wheat may therefore maintain growth in a longer season, even when parts of the season shift to times of the year with less irradiance, which is a potential problem at increasing latitudes (Körner, 2006). The degree of injury of wheat from spring freezes is influenced by the duration of the low temperatures as well as the low point they reach. Prolonged exposure to freezing causes much more injury than brief exposure to the same temperature. Temperatures at which injury can be expected are for two hours of exposure to each temperature. Less injury might be expected from shorter exposure times, while injury might be expected at even somewhat higher temperatures from longer exposure times (Warrick and Miller, 1999). Therefore, when the frost damage incurred was relatively early in the season, giving the crop time to recover from the damage (Whaley, et al., 2004).

Vernalization is defined as acceleration of the ability to flower by a chilling treatment (Chouard 1960). Exposure to temperatures in the vernalization range shortens the vegetative phase and decreases the final leaf number of cereals with a vernalization response (Wang et al. 1995). A vernalization response reduces the risk of winter cereals entering the extremely cold sensitive reproductive growth stage until the danger of low-temperature damage has passed. Vernalization response can occur during seed formation and ripening, germination, and seedling growth (Flood and Halloran 1986). The gradual reduction in low-temperature tolerance of plants stored for long periods at temperatures in the vernalization - cold acclimation range were explained by a shutting off of the low-temperature tolerance genetic machinery once vernalization saturation has been achieved (Fowler et al., 1996), but the ability of cereals to cold acclimate after vernalization saturation and before heading indicates that vernalization saturation does not act as an off switch for low-temperature tolerance genes. The possibility that the vernalization genes have a more subtle regulatory role in controlling the expression of low-temperature tolerance genes cannot, however, be ruled out (Fowler et al., 1996).

The ability of wheat plants to maintain frost tolerance decreases after the vegetative/reproductive transition (Mahfoozi et al., 2001a, b), but mechanisms exist which slow down the rate of phenological development and extend the vegetative phase. These involve a requirement for vernalization (the necessity of going through a certain period of low temperatures) and responses to photoperiod (day-length sensitivity). Winter wheat varieties usually have a greater vernalization requirement and a higher sensitivity to short days than spring wheat varieties, which enable them to remain in the vegetative phase during winter. Spring wheat varieties have no, or a very limited, need for vernalization (Prášil et al., 2004).

The study by Bauer and Black (1990) showed a T_{Lmin} of 0 to $-15°C$ for rapid freezing but temperatures down to $-20°C$ were tolerated with gradual cooling. However, wheat seemingly differs in this respect in its response to extreme heat and cold. Thus, plants develop cold hardiness, but not a tolerance to extreme heat (Drozdov et al., 1984) with heat to tolerance increasing by only 1°C following gradual rather than abrupt warming.

1.5 Solar radiation and precipitation role in wheat growth

Other environmental weather factors that affect the crop growth are the solar radiation and precipitation. The solar radiation is affecting on the crop evapotranspiration, where the increased solar radiation increases transpiration and, for each year there is a point after which further increases in radiation cause a water stress yield loss. While changes in the precipitation affect the amount of water in the soil and hence the yield loss due to the water deficit. Hence, at high precipitation levels, there is sufficient water for the crop and so changes in precipitation have no effect. As precipitation reduces, both the length and severity of the water deficit increases and so each reduction in precipitation causes a progressively larger drop in yield. At very low precipitation, the soil would run out of water during the growing period so that further reductions in precipitation would have less effect. By the reduction of the precipitation and the increasing of the solar radiation during different years, that increase the coefficient of variation, which caused by the differences in water stresses experienced during different years. For low values of solar radiation, the water deficit is small enough that the potential yield is obtained. The increase in the solar radiation does not only increase the water deficit through greater transpiration by the plant, but it also increases the rate of accumulation of biomass, when water and nutrient supply is sufficient. Both potential anthesis biomass and potential grain fill biomass are proportional to solar radiation and so the yield is also proportional (Brooks et al., 2001).

1.6 Wheat growth related to the different soil attributes

Nevertheless, that the temperature is one of the most important environmental factors that affects the crop growth, but there are other factors, which also play an important role for the growth. The soil attributes and its relation with the water supply can affect dramatically on the crop growth and yield. The soil is acting as the water reservoir of the plant. The available water in the soil decreases during the growing season limiting plant growth (Semenov, 2007). Soil water content is one of the essential drivers for biological and chemical soil processes in the unsaturated zone (Kätterer et al., 2006). Different crops may respond differently to water limitation, depending on their water requirements. The different soil characteristics also had different responds to the soil capacity for holding more or less water amount. Therefore, the effect of drought on crops should be characterized not in terms of soil water deficit experienced by crop during the growing season, but by the reduction in grain yield caused by water limitation (Semenov, 2007).

For calculating the soil water content or the soil hydraulic properties, Pedotransfer functions (PTFs) are becoming increasingly popular for estimating hydraulic properties such as soil texture, bulk density, organic matter content, and water retention. The majority of PTFs are completely empirical, although physico-empirical models and fractal theory models have also been developed (Minansy and McBratney, 2000). Schaap (1999) categorized PTFs into three main groups: (1) class PTFs, (2) continuous PTFs, and (3) neural network analysis-derived PTFs. Class PTFs are based on the assumption that similar soils exhibit similar hydraulic properties. Examples are pesented by Carsel and Parrish (1988), Wösten et al.

(1995) and Leij et al., (1996). Continuous PTFs provide continuously varying estimates of hydraulic properties across the textual triangle through linear or nonlinear regression models. Model performance can be improved with additional information such as bulk density, porosity, organic matter content, and water retention points (Rawls and Brakensiek, 1985; Vereecken et al., 1989; Rawls et al., 1992; Williams et al., 1992a). Neural networks have recently been developed to improve the predictions of empirical PTFs (Pachepsky et al., 1996; Schaap and Bouten, 1996; Schaap et al., 1998; Tamari et al., 1996).

1.7 The expected climate change effect on wheat crop

Atmospheric CO_2 levels have been steadily rising during the past century, as a result of fossil-fuel burning and land clearing (Siegenthaler and Sarmiento, 1993). The CO_2 levels at 1990 are about 355 ppm, or 25% higher than the pre-industrial value of 280 ppm (Keeling, 1991). Emissions continue to grow and CO_2 concentrations had increased to over 390 ppm, or 39% above preindustrial levels, by the end of 2010 (IPCC, 2011). Other greenhouse gases (CH_4, CFCs, NO_2) are also on the rise in the atmosphere. If the current rate of emissions continues, global mean temperatures are predicted to increase approximately 1.5°C by the middle of the century (Hansen et al., 1988; Houghton et al., 1990), and the temperature increase of the 100-year linear trend (1906-2005) was 0.74 [0.56 to 0.92]°C (IPCC, 2007). Precipitation patterns are also expected to change. Many studies indicate that high CO_2 levels and rising mean temperatures will affect crop yields (Kimball, 1983; Acock & Allen, 1985; Acock, 1991). While CO_2 alone would most probably increase yields, interactions with factors like temperature, precipitation and management practices make predictions less certain (Tubiello et al., 1995). Changes to the global climate, notably to regional spatial and temporal temperature patterns (Houghton et al., 1996), from increased atmospheric concentrations of greenhouse gases are predicted to have important consequences for crop production (Parry, 1990). Changes in climate may exhibit increased climatic variability and small changes in climatic variability can produce relatively large changes in the frequency of extreme climatic events (Kattenberg et al., 1996).

1.8 The effect of the expected temperature increase on wheat growth

Wheeler et al. (2000) found that increasing the mean seasonal temperature by 2°C will decrease the grain yield by 7%, and the rapid decline in grain yields was associated with a reduction in the number of grains per year at the time of harvest maturity. The negative effect of warmer temperatures should be countered by the increased rate of crop growth at elevated atmospheric CO_2 concentrations at least when there is sufficient water. Of more importance for the yield of annual seed crops may be changes in the frequency of hot (or cold) temperatures which are associated with warmer mean climates. Seed yields are particularly sensitive to brief episodes of hot temperatures if these coincide with critical stages of crop development. Hot temperatures at the time of flowering can reduce the potential number of seeds or grains that subsequently contribute to the crop yield. Therefore, with an increase in mean temperature, winter wheat is expected to resume growth earlier and spring wheat to be sown earlier, and the crops would develop under much the same thermal environments as at present (Batts et al., 1997). In addition to the occurrence of high temperature could have an impact at vulnerable stages and be much more damaging to crop yields (Semenov and Porter, 1995; Nicholls, 1997; Porter and Gawith, 1999; Porter and Semenov, 1999) than more stable conditions. As well, temperature sensitivity and thus responsiveness to extreme temperature events vary during the course of crop development (Slafer and Rawson, 1994, 1995c).

Winter wheat is generally sown world-wide in sub-optimal temperatures, which fluctuate between 8 to 16°C mean daily temperature, and at the warmer conditions, which will result from the climate change, during sowing should not have a negative impact on wheat establishment (Porter and Gawith, 1999). Nevertheless, if temperature increases as predicted at the grain-filling stage, then wheat can mature earlier, with different time depending on the supposed climate scenario, compared with the baseline climate. This has the following consequences. First, the duration of the grain filling stage, measured in calendar days, decreases resulting in the lower amount of radiation intercepted by the plant during grain filling. This, in turn, reduces production of new biomass during grain filling decreasing the final yield. Secondly, the grain filling stage will occur early in a season, when expected daily radiation is sub-optimal, i.e. lower, on average. This reduces grain yield even further. As a result, wheat yield decreases with global warming, if other factors are taken out of consideration (Semenov, 2007).

1.9 The direct effect of the expected increase of CO_2

There is a large, but not complete, agreement between climatologists that the global mean temperature in mid-latitudes is increasing due to an increase in the concentration of atmospheric CO_2 and other greenhouse gases (IPCC report 1990; cf. Idso, 1989). Historical and modern records show that the atmospheric carbon dioxide (CO_2) concentration increased from approximately 280 ppm in pre-industrial times to about 315 ppm by 1958, and to more than 350 ppm by 1988 (Boden et al., 1994), and had increased to over 390 ppm, or 39% above preindustrial levels, by the end of 2010 (IPCC, 2011). The accelerated trend in the global CO_2 growth rate during the first 30 years of modern records has led to various scenarios for the future CO_2 concentrations of the atmosphere (Hunsaker et al., 2000). Along with changes in temperatures and shifts in precipitation pattern, periods of drought are predicted to increase in the future (IPCC, 2001). While large uncertainties remain about the extent of these climate changes at the regional scale, increasing atmospheric concentration of CO_2 is among the most predictable aspects of global environmental change.

Elevated CO_2 causes partial stomatal closure thereby decreasing leaf transpiration at the same time that carbon assimilation is increased (Morison, 1998). Therefore, because soil water availability is a major environmental factor for plant growth, it is of particular importance to analyse possible interactions of elevated CO_2 and water supply in terms of the water use of plant canopies (Burkart et al., 2004). Responses of canopy evapotranspiration (EC) to elevated CO_2 have been found to vary from positive (Chaudhuri et al., 1990; Kimball et al., 1994; Hui et al., 2001), to unchanged (Jones et al., 1985; Hileman et al., 1994; Ellsworth, 1999) to negative (Ham et al., 1995; Drake et al., 1997; Kimball et al., 1999). The variability of these responses may result from several feedback mechanisms that effect EC, including irradiance, wind speed, canopy temperature, VPD, leaf conductance, canopy leaf area (Kimball et al., 1994; Wilson et al., 1999). It is in fact well-recognized that CO_2 concentration and management factors will interact in complex ways to determine the ultimate impacts of climate change on crop production. While elevated CO_2 alone tends to increase growth and yield of most agricultural plants (Kimball, 1983; Cure and Acock, 1986; Allen et al., 1997; Kimball et al., 2002), warmer temperatures and changed precipitation regimes may either benefit or damage agricultural systems (e.g., Rosenzweig and Hillel, 1998). Water and fertilizer application regimes will further modify crop responses to elevated CO_2 (e.g., Reilly et al., 2001).

Rainfed crops were found to be more sensitive to CO_2 increases than irrigated ones. On the other hand, low nitrogen applications depressed the ability of the wheat crop to respond positively to CO_2 increases. In general, the positive effects of high CO_2 on grain yield were found to be almost completely counterbalanced by the negative effects of high temperatures.

Depending on how temperature minima and maxima were increased, yield changes averaged across management practices ranged from -4% to 8% (Tubiello et al., 1995).

On the other hand, elevated CO_2 had only minor effects on growth and plant performance, but positive effects on root biomass in autumn and frost tolerance in November. It had been found that elevated CO_2 improved plant growth from sowing only up until the first harvest; a period with mean temperatures of 9-12°C. Later in the autumn there were no growth responses of shoots to elevated CO_2 (Morison & Lawlor, 1999).

1.10 Wheat crop models

Crop models are defined as dynamic representations of crop processes in a systems context. The goal of such models is to simulate and explain crop development and behavior, yield and quality as a function of environmental and management conditions or of genetic variation. The explicit processes or mechanisms that are the basis of such crop models warrant calling them process or mechanistic models in contrast to the `black-box' statistical models. The statistical models, used to empirically predict large-scale (county to region) agricultural yields from regression analyses based on monthly or annual variables; and process oriented models, further referred to as process models, used to compute crop dynamics at smaller spatial scales (leaf to canopy and/or field levels), based on deterministic equations and simulation of underlying processes at timescales of minutes to days. Process models can be further grouped into 'complex' and 'simple'. Complex models compute processes at the level of organs or lower; for example, the dynamics of carbon and water calculated at the leaf-level, requiring time-steps ranging from minutes to hours. Simple models compute canopy-level dynamics directly, using empirical relationships without consideration of underlying processes, typically using daily time steps (Passioura, 1979; Thornley, 1980; Charles-Edwards et al., 1986; Sinclair and Seligman, 2000).

In general, both statistical and process models adequately predict agronomic yields at given scales. Statistical models were intrinsically designed to operate at the multi-seasonal, regional scale, and are thus best suited for analyzing interannual variability of regional production. Process crop models were developed to simulate crop responses to environmental conditions at the plot and field level and can be used to analyze interseasonal dynamics of field-level crops. Process crop models capture the dynamics of crop response to elevated CO_2, and therefore have been widely used in climate change studies. These models have different components that can be simplistically grouped into those computing: plant phenology as a function of accumulated temperature and daylength; photosynthesis and respiration; water balance, N-uptake and distribution and effects of other factors; partitioning, biomass accumulation and organ growth. These components may operate at different timescales. For instance, photosynthesis and water exchange are resolved at timescales from minutes to hours (complex process models) to days (simple process models). Biomass production and partitioning, and ultimately yield, are generally computed at daily (process models) to seasonal (statistical models) time-steps. Thus, linkages among model components are often across timescales (Tubiello and Ewart, 2002).

The aim of this study is developing a new wheat crop simulation model, which can simulate the winter wheat crop performance (yield, biomass, development stages timing) affected by the different environmental factors at different spatial and temporal scales at the current time and in the future. The developed model here is classified as a simple process model, which is not so complicated as the complex process models or not so very simple as the statistical models, where the main advantages of a simpler model are: a) easier to understand and consequently easier to interpret the results from the model; b) quicker to build, test and run (Brooks and Tobias, 1996). Nevertheless, the main disadvantage with using a simple model is that important aspects of the system may be omitted, so that the model is unrealistic (Brooks et al., 2001).

Wheat simulation models are being used increasingly to assess the crop performance according to different environmental factors such as weather factors, soil attributes, and water supply, which could be optimum or not for obtaining the maximum yield. These types of applications have been built for monitoring the important results such as the grain yield, yield variability, and geographical distribution of the crop, affected by the environmental factors at present and in the future. The crop simulation model could also predict the timing of crop development because first, the effects of environment (temperature and photoperiod) on crop development (phenology) are central to crop adaptation (Evans, 1993; Roberts et al., 1996) and second, the accuracy of the phenology sub-model of a crop simulation model is a sensitive step to the precise prediction of crop biomass and yield (Wheeler et al., 2000).

Thermal-germination models usually include some assumptions regarding within-population variability in growth rate response, from germination to grain-filling stages, to temperature (Hardegree et al., 1999). These assumptions are most often related to cardinal-temperature (CT) concepts such as the base temperature (T_{base}), optimal temperature (T_{opt}), maximum or ceiling temperature (T_{max}), and sub-optimal and supra-optimal thermal time (θ_1, θ_2) (Garcia-Huidobro et al., 1982a, b; Covell et al., 1986; Ellis et al., 1986; Ellis and Butcher, 1988; Roberts, 1988; Benech Arnold et al., 1990; Probert, 1992; Alvarado and Bradford, 2002). CT variables are typically assigned either a constant value, or are determined to be normally or log-normally distributed within a given seed population (Covell et al., 1986).

The response of yield is the most sensitive to temperature. The main effect of temperature is to set the phenological dates and so changes in temperature affect both the timing and duration of the main growth periods in which most of the biomass is accumulated and during which the water deficit tends to increase (Brooks et al., 2001). The reasons of predicting the timing of crop development are: First, the effects of environment (temperature and photoperiod) on crop development (phenology) are central to crop adaptation (Evans, 1993; Roberts et al., 1996). Second, the accuracy of the phenology sub-model of a crop simulation model is a sensitive step to the precise prediction of crop biomass and yield (Wheeler et al., 2000). Nevertheless, the crop simulation models use this strategy for distributing the accumulated biomass at each phenological stage. And according to the annual variability of temperature, the annual biomass and yield vary too. The major methods have been used in models to simulate the timing of important events in lifecycle in wheat, and the influence of daylength and vernalisation on their timing. The older of these methods (e.g., Ritchie and Otter, 1985; Weir et al., 1984) describe development through sequential phases (e.g., floral initiation, terminal spikelet, flag leaf appearance) leading to anthesis. Intervals of modified thermal time are assumed to be constant for these phases, and key stages such as anthesis are reached when the sequence of phases has been completed (Weir et al., 1984; Ritchie and Otter, 1985). The modification of thermal time is achieved by slowing its accumulation by multiplying the daily increment of thermal time by daylength and vernalisation factors (values 0 – 1). The factors are applied during particular phases, and vernalisation and daylength responses are completed once an apical stage is reached where the factor is no longer appropriate. The daylength and vernalisation factors are applied either as most limiting (Ritchie and Otter, 1985) or factorially (Weir et al., 1984). A variation on this approach was to use thermal time either directly or as represented by the number of leaves between developmental events (McMaster et al., 1992). Accurate simulation of anthesis using this method, particularly in winter wheat, is dependent on accurately simulating earlier development events, particularly the switch at the shoot apex that determines whether the primordial will differentiate into leaves or spikelets, or floral initiation (Jamieson et al., 2007).

Wheat simulation models such as Sirius (Jamieson et al., 1998c), AFRCWHEAT2 (Weir et al., 1984; Porter, 1993), CERES-Wheat (Ritchie and Otter, 1985), and ECOSYS (van Laar et al., 1997) are designed to simulate the growth and development of wheat in small, homogeneous areas. They require input data for weather, soil attributes and management practice (choice of cultivar, sowing date, nitrogen application and irrigation) at varying detail. They are able to supply output, on daily basis, of variables such as biomass, yield, soil water content, mainstem leaf number, leaf area and evapotranspiration (Brooks et al., 2001).

1.11 Climate prediction

Many crop simulation models require daily site-specific weather as their input (Jamieson et al., 1998b). To use process-based models for the assessment of impacts of climate not only at the current conditions, but also under the climate change effect. Climate scenarios have to be developed with appropriate temporal and spatial resolutions, taking into account the model sensitivity to variations in climatic variables. Crop simulation models incorporate a mixture of non-linear interactions between the crop and its environment (Porter and Semenov, 1999, 2005; Semenov and Porter, 1995). Those non-linear models can potentially produce very different predictions depending on how climate scenarios were constructed (Mearns et al., 1997). It was demonstrated by Porter and Semenov (1999) and Semenov and Barrow (1997) that climate change scenarios, derived from a global climate model (GCM) that incorporated changes in climatic variability decreased mean wheat yield and significantly increased risk of crop failure compared with scenarios which accounted only for changes in mean values. Katz and Brown (1992) analysed the sensitivity of weather extreme events to changes in the mean and variability of climatic variables, and found that extreme events are more sensitive to changes in variability than to changes in the average. But the coarse spatial resolution of GCMs and large uncertainty in their output at a daily scale, particularly for precipitation, means that the daily GCM output is not appropriate directly for use with process-based models and analysis of extreme events (Trigo and Palutikof, 2001). Hence, there are two methods for overcoming the uncertainty of the GCMs. First, and it is the most common used way, is applying the predicted changes in climate models to the weather generator parameters (Semenov and Brooks, 1999). A stochastic weather generator is a numerical model which produces synthetic daily time series of a suite of climate variables, such as precipitation, temperature and solar radiation, with certain statistical properties (Richardson, 1981; Richardson and Wright, 1984; Racsko et al., 1991). Weather generators were adopted in climate change impact studies as a computationally inexpensive tool to generate scenarios with high temporal and spatial resolutions based on the output from a global climate model (GCM) (Wilks, 1992; Barrow and Semenov, 1995; Wilks and Wilby, 1999; Hansen, 2002; Dubrovsky et al., 2005). These statistical approaches rely on the existence of a long historical observational record from which statistical relationships can be calculated. In practice, this limits the variables that can be downscaled to generally temperature and rainfall, and the location of observation station determines where the downscaling can be applied. There is another important method that also depends on the downscaling of the GCM outputs called Regional Climate Models (RCM). Dynamical downscaling, or the use of Regional Climate Models (RCMs), is often seen as the alternative to statistical downscaling though most statistical approaches also can be applied to RCM output to obtain point location data (Evans, 2012). They include dynamic downscaling, which uses complex algorithms at a fine grid-scale describing atmospheric process nested within the GCM outputs (Jones et al., 1995). RCMs have some advantages over statistical techniques: They simulate the entire climate system so that all climate variables of interest are available, rather than being limited to the well observed variables; and they simulate the climate across the landscape regardless of whether observations exist (Evans, 2012). Second, instead of downscaling output from seasonal weather forecast, using a stochastic weather generator, it is possible to upscale a crop model to one, which can operate on a larger regional scale and can take seasonal predictions from GCM output directly without the need for spatial and temporal downscaling (Challinor et al., 2004).

2 Materials and Methods

2.1 Experimental data set

A growth development model has been designed to predict the biomass and grain yield of wheat crop in Bavaria. The plant, soil, and weather data were collected from the experimental site Dürnast of the Chair of Plant Nutrition, Technische Universtät München for Freising, and other eleven locations from the experimental sites of LfL (Bavarian State Research Center). The experimental data set was gathered for the seasons' duration from 2000/01 to 2008/09. The experimental sites for the latter studied ones had sometimes no nearby weather stations; therefore, the weather data had been taken from the nearest location to the experimental studied area (Table 1).

Table1: The details of the studied experimental sites and its distance to the used weather stations.

Experiment location	Altitude (m a.s.l)	Nearest weather station	Distance (km)
Würzburg - Giebelstadt	430	Veitshöchheim	25
Donau-Ries – Reimlingen	542	Wallerstein	10
Freising - Dürnast	290	–	–
Passau - Reith	360	–	–
Regensburg - Köfering	350	–	–
Lichtenfels - Wolfsdorf	295	–	–
Eichstätt – Desching	530	Hepberg	5
Landshut – Feistenaich	370	–	–
Neumarkt – Hartenhof	460	Sommertshof	10
Weissburg-Gunzenhausen – Bieswang	280	Windsfeld	30
Main-Spessart – Arnstein	470	Schweinfurt – Ettleben	13
Günzburg - Günzburg	470	Weißingen	12

Table2: The soil type and the planting date for the studied seasons at the studied sites

Season		Freising	Reimlingen	Wolfsdorf	Reith	Köfering	Giebelstadt	Desching	Feistenaich	Hartenhof	Bieswang	Arnstein	Günzburg
2000/01	Soil type	Silt loam	Loam	Sandy loam	Silt loam	Silt loam	Silt loam	Loam	Silt loam	Sandy loam	Clay loam	Silt loam	Silt loam
	Planting date	20.10	18.10	19.10	30.10	16.10	16.10	19.10	17.10	18.10	23.10	25.9	16.10
2001/02	Soil type	-	Loam	Sandy loam	Silt loam	Silt loam	Silt loam	Clay loam	Silt loam	Sandy loam	Clay loam	Clay loam	Silt loam
	Planting date	-	16.10	23.10	30.10	23.10	16.10	19.10	16.10	17.10	19.10	16.10	16.10
2002/03	Soil type	Silt loam	Loam	Sandy loam	Silt loam	Loam	Silt loam	Sandy loam	Silt loam	Loam	Loam	Clay loam	-
	Planting date	22.10	9.12	10.10	29.10	14.10	11.10	10.10	11.10	6.11	22.10	10.10	-
2003/04	Soil type	Silt loam	Loam	Loam	Silt loam	Loam	Silt loam	Loam	Silt loam	Loam	Loam	Clay loam	Silt loam
	Planting date	14.10	13.10	15.10	16.10	14.10	13.10	2.10	30.10	15.10	14.10	10.10	15.10
2004/05	Soil type	Silt loam	Loam	Loam	Silt loam	Loam	Silt loam	Loam	Silt loam	Loam	Loam	Clay loam	Silt loam
	Planting date	13.10	13.10	15.10	16.10	14.10	15.10	2.10	30.10	15.10	14.10	10.10	15.10
2005/06	Soil type	Silt loam	Loam	Sandy loam	Silt loam	Loam	Silt loam	Loam	Silt loam	-	Loam	Clay loam	Silt loam
	Planting date	11.10	12.10	13.10	26.10	11.10	10.10	14.10	20.10	-	17.10	17.10	13.10
2006/07	Soil type	Silt loam	Loam	Sandy loam	Silt loam	Silt loam	Silt loam	Loam	Silt loam	Sandy loam	Loam	Clay loam	Silt loam
	Planting date	10.10	5.10	16.10	17.10	13.10	16.10	12.10	11.10	16.10	11.10	10.10	11.10
2007/08	Soil type	Silt loam	Loam	Sandy loam	Silt loam	Silt loam	Silt loam	-	Silt loam	Sandy loam	-	Clay loam	Silt loam
	Planting date	11.10	8.10	12.10	17.10	15.10	19.10	-	12.10	16.10	-	9.10	12.10
2008/09	Soil type	Silt loam	Loam	Sandy loam	Silt loam	Silt loam	Silt loam	-	Silt loam	Sandy loam	Loam	Clay loam	Silt loam
	Planting date	14.10	10.10	15.10	21.10	9.10	21.10	-	8.10	13.10	14.10	15.10	10.10

Daily weather data were used from the selected weather stations, where the used weather elements were: maximum and minimum temperature (°C), mean relative humidity (%), precipitation (mm), solar radiation (MJ/m^2.d^1), and wind speed (m/s)   eather data were collected from the LfL´s official website (http://www.wetter-by.de). The experimental sites seasons had different soil types, which include; loam, sandy, clay loam, and silt loam. the yield of these sites during the studied seasons was taken for comparison, plus the planting date, and the soil type to have more precise simulation. Table 2 shows the detailed plant data, which have been taken into consideration for the simulation.

2.2 The methodology of the wheat crop model

The model was set up to evaluate the biomass and grain yield of a wheat growth season, affected by different environmental factors (weather data, soil attributes, and water supply). It describes the partitioning of the biomass during the different phenological stages. This model uses the weather data, water supply and soil attributes for each selected location as an input data.

2.3 Evapotranspiration and Soil Water Flow

2.3.1 Potential Evapotranspiration

The calculation steps of the model start first from the effect of the external factors on the growth. Starting from the effect of the water supply (surplus or deficit) on the plant, the potential and crop evapotranspiration factors for the wheat crop have been calculated by using the FAO Penman-Monteith equations (Allen, 1998, 2000). The potential evapotranspiration ET_0 [mm day^{-1}] is calculated as a daily factor, which is based mainly on the weather data:

$$ET_0 = \frac{0.408\Delta(R_n - G) + \gamma \frac{900}{T+273} U_2 (e_s - e_a)}{\Delta + \gamma(1 + 0.34 U_2)} \tag{1}$$

where, R_n is net radiation at the crop surface [MJ m^{-2} day^{-1}], G is soil heat flux density [MJ m^{-2} day^{-1}] and could be ignored, T is mean daily air temperature at 2 m height [°C], U_2 is wind speed at 2 m height [m s^{-1}], e_s is saturation vapour pressure [kPa], e_a is actual vapour pressure [kPa], Δ is slope vapour pressure curve [kPa °C^{-1}], and γ is psychrometric constant [kPa °C^{-1}].

The last parameters have been also calculated by using the daily maximum, minimum and average temperature T_{max}, T_{min}, and T_{avg} respectively [°C], relative humidity RH [%], atmospheric pressure P_a [kPa], the saturated vapor pressure $e°T_{max}$ at the maximum and $e°T_{min}$ at minimum temperature [kPa], and λ is the latent heat vaporization [MJ kg^{-1}], using the following equations:

$$e_s = \frac{e°T_{max} + e°T_{min}}{2} \tag{2}$$

$$e°T_{max} = 0.6108 \times exp\left(\frac{T_{max} \times 17.27}{T_{max} + 273.15}\right) \tag{3}$$

$$e°T_{min} = 0.6108 \times exp\left(\frac{T_{min} \times 17.27}{T_{min} + 273.15}\right) \tag{4}$$

$$e_a = \frac{RH}{100} \times e_s \tag{5}$$

$$\Delta = \frac{4098 \times \left[0.6108 \times exp\left(\frac{T_{avg} \times 17.27}{T_{avg} + 273.15}\right)\right]}{(T_{avg} + 273.15)^2} \tag{6}$$

$$\gamma = \frac{1.013 \times 10^{-3} \times P_a}{0.622 \times \lambda} \tag{7}$$

$$\lambda = 2.501 - (2.361 \times 10^{-3}) \times T_{avg} \tag{8}$$

2.3.2 Actual Transpiration

The crop evapotranspiration (ET_c) has been calculated by multiplying the basal crop coefficient (K_{cb}), the soil water evaporation coefficient (K_e) and the transpiration coefficient (K_s) with the potential evapotranspiration (ET_0) (Allen, 1998) as follows:

$$ET_c = (K_{cb} \times K_S + K_e)ET_o \tag{9}$$

The K_{cb} value has been retrieved from Allen (1998), depending on the crop type and its present development stage (Dooernbos and Kassam, 1979; Doorenbos and Pruitt, 1977; Snyder et al., 1989), while the other two coefficients are calculated with the following equations:

$$K_s = min(K_r(K_{c,max} - K_{cb}), f_{ew} \times K_{c,max}) \tag{10}$$

where, $K_{c,max}$ is the upper limit of the evaporation and transpiration from any cropped surface, f_{ew} is the fraction of soil surface from which most evaporation occurs, and K_r represents a dimensionless evaporation reduction coefficient, which is dependent on the cumulative depth of water depleted (evaporated) from the topsoil. K_s is the soil water evaporation coefficient, which represents the soil water status, affected by the soil type, precipitation frequency, weather data, crop species and its development stage.

$$K_{c,max} = \left(\left\{1.2 + [0.04(U_2 - 2) - 0.004 \times (RH_{min} - 45)] \times \left(\frac{pt}{3}\right)^{0.3}\right\}, \{K_{cb} + 0.05\}\right) \quad (11)$$

where, U_2 is wind speed at 2 m height [m s^{-1}], and pt is the plant height [m]. The plant height value is retrieved from Penning de Vries et al., (1989), depending on the crop type and its present developmental stage.

$$f_{ew} = min(1 - f_c, f_w) \quad (12)$$

where, (1 - f_c) is the average exposed soil fraction not covered (or shaded) by vegetation, (0 ≤ f_c ≤ 0.99) and f_w (0 ≤ f_w ≤ 1) is the average fraction of soil surface wetted by irrigation or precipitation which in case of precipitation is equal to 1. The effective fraction fc of soil surface covered by vegetation is given by

$$f_c = \left(\frac{k_{cb} - k_{c,min}}{k_{c,max} - k_{c,min}}\right)^{(1+0.5 \times pt)} \quad (13)$$

where $k_{c,min}$ is the minimum value of the crop coefficient (K_c) for dry bare soil with no ground cover and takes values between 0.15 and 0.20 [dimensionless], where K_c equals:

$$K_c = K_{cb} + K_e \quad (14)$$

The reduction coefficient of evaporation is calculated by

$$K_r = \frac{(TEW - D_{e,i-1})}{(TEW - REW)} \quad (15)$$

where, TEW is the maximum cumulative depth of evaporation (depletion) from the soil surface layer when K_r = 0 (TEW = total evaporable water) [mm], REW is the cumulative depth of evaporation (depletion) at the end of stage 1 (REW = readily evaporable water) [mm], and D_{ei} is the cumulative depth of evaporation (depletion) from the soil surface layer at the end of day i-1 (the previous day) [mm].

2.3.3 Soil water flow

TEW and REW are soil parameters, which are dependent on available water in the soil, and are calculated as follows:

$$TEW = 10(\theta_{FC} - 0.5\,\theta_{WP})Z_e \quad (16)$$

$$REW = (3.121\,TEW + 22.896)Z_e \quad (17)$$

where, θ_{FC} and θ_{WP} is the volumetric percentage of the soil water at field capacity and wilting point respectively, and Z_e is the depth of the soil surface soil layer that is subject to drying by way of evaporation, which takes values between 0.10 and 0.15 m.

$$D_{e,i} = D_{e,i-1} - (P_i - RO_i) - \frac{I_i}{f_w} - \frac{E_i}{f_{ew}} + DP_{e,i} \quad (18)$$

where, P_i is the daily precipitation amount [mm] on day i, RO_i is precipitation run off from the soil surface [mm], which is assumed to be a constant percentage from the precipitation amount. In case of irrigation the fraction I_i/f_w is describing the infiltration through the wetted part of the soil surface represented by f_w, where I_i is the irrigation depth for the part of the surface that is wetted [mm], and $DP_{e,i}$ is the deep percolation loss from the topsoil layer on day i if soil water content exceeds field capacity [mm], which depends also on the daily precipitation and its soil surface run off and the cumulative depth of evaporation from the soil surface for the day i-1.

$$DP_{e,i} = (P_i - RO_i) - \frac{I_i}{f_w} - D_{e,i-1} \qquad \geq 0 \qquad (19)$$

The third ET_c dependant coefficient is the transpiration reduction coefficient (K_s), which represents the soil water availability in the root zone and the transpiration amount of it

$$K_s = \left[\left(1 - (b/k_y \times 100)\right) \times \left(Ec_e - Ec_{e,threshold}\right)\right] \times \left[(TAW - D_{r,i})/(TAW - RAW)\right] \qquad (20a)$$

This equation is applied only if $D_{r,i} > RAW$, but if $D_{r,i} \leq RAW$, then the transpiration reduction coefficient (K_s) will be

$$K_s = \left[\left(1 - (b/k_y \times 100)\right) \times \left(Ec_e - Ec_{e,threshold}\right)\right] \qquad (20b)$$

where, TAW is the total available soil water in root zone [mm], which represents the available soil water content compared with the present root zone. RAW is the readily available soil water in the root zone [mm], which equals TAW multiplying by average fraction of Total Available Soil Water (TAW) that can be depleted from the root zone before moisture stress (reduction in ET) occurs ($0 \leq TAW \leq 1$). $D_{r,i}$ is the root zone depletion at the end of day i.

$$TAW = 1000(\theta_{FC} - \theta_{WP})Z_r \qquad (21)$$

$$RAW = (p + 0.04(5 - ET_c))TAW \qquad (22)$$

where, p is an average fraction of Total Available Soil Water (TAW) that can be depleted from the root zone before moisture stress (reduction in ET) occurs [0 - 1] and it equals 0.55 in case of winter wheat. And for calculating the root zone depletion D_r, there is two equations, the first one is used only for initialisation

$$D_{r,i} = 1000(\theta_{FC} - \theta_{i-1})Z_r \qquad (23a)$$

where, θ is the average soil water content for the effective root zone at day (i-1). The second equation is applied for the rest of the growth

$$D_{r,i} = D_{r,i-1} - (P - RO)_i - I_i + ET_{c,i} + DP_i \qquad (23b)$$

For calculating the soil water content at field capacity and wilting point, the van Genuchten equation has been used (van Genuchten, 1980) as follow:

$$\theta = \theta_r + (\theta_s + \theta_r)[1 + (\alpha|h|)^n]^{-m} \qquad (24)$$

where, h is the pressure head [cm], and it is assumed to be -15000 cm at the wilting point and -330 cm at field capacity for calculating the volumetric soil water θ at the wilting point and field capacity respectively, θ_r and θ_s are the residual and saturated volumetric water content respectively [cm^3 cm^{-3}], and α, n, m are the moisture retention characteristic parameters. The last unknown parameters have been calculated from the pedotransfer function of Vereecken et al, (1989), according to Priesack (2006).

$$\theta_s = 0.81 - 0.28\,\rho_s + 0.13\,f_{clay} \qquad (25)$$

$$\theta_r = 0.015 + 0.5\,f_{clay} + 1.39\,f_{Corg} \qquad (26)$$

$$\log(\alpha) = -2.49 + 2.5\,f_{sand} - 35.1\,f_{Corg} - 2.62\,\rho_s - 2.3\,f_{clay} \qquad (27)$$

$$\log(n) = 0.05 - 0.9\,f_{sand} - 1.3\,f_{clay} + 1.5\,f_{sand}^2 \qquad (28)$$

where, f_{sand}, f_{clay}, f_{Corg} are the fractions of the sand, clay and organic carbon respectively, in the soil, and ρ_s is the bulk density [g cm^{-3}] and m=1.

The calculated ET$_c$ has been then used for identifying the occurrence of water deficit or water surplus in the soil by another set of equations by Brader (1986). These equations aim to calculate the water requirement satisfaction index (WRSI), which indicates the extent to which the water requirements of the annual crops have been satisfied in a cumulative way at any stage of the crop growing season. The calculations of the WRSI start from the calculation of the difference between the actual precipitation (P) and crop evapotranspiration (ET$_c$) for each 10 days (P_E$_{10}$). This value could be positive or negative, depending on the accumulated precipitation is higher or lower than the accumulated ET$_c$. Then the water reserves in the soil (R$_{s,i}$), which express the quantity of water existing within the rooting zone of the crop, depending on the soil type and depth, are calculated by multiplying the difference between the soil water content at field capacity and wilting point with the useful depth of the soil profile exploited by the crop's root. This value of R$_{s,i}$ is an initial value, which is changed by adding the P_E$_{10}$ to it, taking into consideration not to be higher than the soil available water content (AWC) or lower than zero.

$$AWC = 1000(\theta_{FC} - \theta_{WP})Z_r \qquad (29)$$

$$R_{s,i(int)} = AWC \qquad (30a)$$

$$R_{s,i} = P\_E_{10} + R_{s,i-1} \qquad AWC \geq R_{s,i} \geq 0 \qquad (30b)$$

The difference between the calculated $R_{s,i}$ with and without the last condition ($AWC \geq R_s \geq 0$) is representing the surpluses or deficits of soil water (S_D). WRSI in the case of no deficit will be 100, which corresponds to the absence of yield reduction caused by water deficit. In the case of water deficit S_D is less than zero, and according to its value, the WRSI value will decrease in turn.

$$Int_i = Int_{i-1} + \left(\frac{S\_D \times 100}{ET_c \times S\_D}\right) \qquad (31)$$

$$WRSI = 100 - Int_i \qquad (32)$$

The water requirement satisfaction index (WRSI) is an indicator of crop performance based on the availability of water to the crop for each decade during the season. Following Brader (1986) the WRSI values were scaled against the expected decrease percentage of yield (%Y), and from that scale, an estimated equation was developed for calculating the %Y.

$$\%Y = a + b \Big/ \left(1 + \left(\frac{WRSI-c}{d}\right)^2\right) \qquad 100 \geq WRSI \geq 0 \qquad (33)$$

where, the developed parameters a, b, c, and d equal -37.6, 231.1, 136,2, and 43.9 respectively. %Y value has been used directly in the dry matter accumulation equations.

2.4 Plant Growth

2.4.1 Phenological Development

The daily dry matter accumulation operation of the plant depends mainly on the temperature. The growth affecting temperatures are the base temperature for each development stage, where the growth gets in that development stage after reaching that temperature; and the accumulated thermal units, which sets the duration for each development stage. According to the different speed of the accumulated thermal units, the dry matter will differ.

$$RT_i = \left(\frac{T_{max,i}+T_{min,i}}{2}\right) - T_{base} \qquad \geq 0 \qquad (34)$$

$$RT_{sum,i} = RT_{sum,i-1} + RT_i \qquad (35)$$

where, RT is the relative temperature increase during the season (°C), and this value is represented by the subtraction of the average daily temperature (represented as the maximum + the minimum temperature divided by 2) from the base temperature (T_{base}) for each phenological stage. In case RT_i has a minus value at day i, it will be set to zero indicating no growth at day i. The RT_{sum} is the sum of the accumulated daily RT value for calculating the thermal units for each development stage. Table 3 shows the assumed base temperature and thermal units for each developmental stage (Wilsie, 1962; Petr, 1991; Slafer and Savin, 1991; Russell and Wilson, 1994; Rawson and Macpherson, 2000; McMaster et al., 2008).

Table 3. The base temperature (°C) and the accumulated thermal units (°Cd) for each developmental stage during the winter wheat crop season.

Development stage	Sowing	Emergence	Double-ridge	Terminal spikelet	Anthesis	Grain-filling
Base temperature (°C)	3	5	5	2	9.5	7
Thermal units (°Cd)	60	200	150	400	100	590

2.4.2 Crop Growth

The crop growth steps have been mainly calculated by using the equations of Goudriaan and van Laar (1994) from equation 36 to 49.

$$CGR_i = (BM_i - BM_{i-1}) - (age_i - age_{i-1}) \qquad (36)$$

$$RWLV_i = FLV \times CGR_i \qquad (37)$$

The CGR_i is the crop growth rate at day i, which equals the biomass BM [g.dm m^{-2} d^{-1}] for only the day i, where the initiative value for CGR was assumed to be 0. RWLV is the growth rate of leaf biomass, and FLV represents the partitioning fraction of the leaf, which its value has been taken from Penning de Vries et al, (1989) depending on the current development stage.

$$RCR_{RES} = 2 \times TGTB \qquad (38)$$

where, RCR_{RES} is the relative consumption rate of the assimilate reserves and TGTB is a temperature dependent variable (Goudriaan and van Laar, 1994).

Table 4. The relative temperature effect on relative consumption rate of the assimilate reserves (TGTB) values at different temperatures.

Temperature (°C)	0	8	30	40
TGTB	0	0	1	1

$$GR_{SM} = RCR_{RES} \times RES_{i-1} \qquad (39)$$

$$G_{rsp} = \varphi \times GR_{SM} \qquad (40)$$

$$CR = G_f \times GR_{SM} \qquad (41)$$

where, GR_{SM} is the growth rate of structural dry matter [g.dm m^{-2} d^{-1}], G_{rsp} is the growth respiration [g.CO_2 m^{-2} d^{-1}], CR is the consumption rate of assimilation [g.CH_2O m^{-2} d^{-1}]. The factors φ is the CO_2 production factor and equals 0.431 for winter wheat, and G_f is the glucose requirement factor and equals 2.452 for winter wheat (Penning de Vries et al., 1983).

The dry matter is then accumulated daily in the structural biomass (S_{BM}) [g.dm m^{-2} d^{-1}] and the assimilate pools (RES) [g.dm m^{-2} d^{-1}]. The structural biomass is the storage of the assimilates in a reserve pool as a short-term pool for continuing the growth processes and so the conversion of the stored assimilates. While the assimilate pool (RES) [g.dm m^{-2} d^{-1}] is a long-term pool, which serves to store materials during a longer period for subsequent filling of the grain or the formation of new tillers

$$S_{BM,i} = S_{BM,i-1} + GR_{SM} \tag{42}$$

$$MT = 0.014 \times BM \tag{43}$$

$$AS = 45 \times AMTMPT \tag{44}$$

$$R_{RES} = AS - CR - MT \tag{45}$$

$$RES_i = RES_{i-1} + R_{RES} \tag{46}$$

where, AS is the assimilate flow and this value has been calculated from a function depending basically on the temperature (AMTMPT) (Goudriaan and Van Laar, 1994), MT is the daily maintenance [g.CH_2O m^{-2} d^{-1}] and it is assumed to be 0.014 from the biomass value.

2.4.3 Environmental Factors

External events affecting directly on the two pools S_{BM} and RES. These external events are; 1) the water deficit effect on the growth %Y, and this factor affects the growth during the whole season except for the last accumulated 100 °Cd, where the water deficit has no more effect on the growth, 2) the extreme temperature effect (x_T) on the anthesis and grain-filling developmental stages. The hot days, which exceed 31°C as a maximum temperature, affect negatively on the growth (Wheeler et al., 2000), by multiplying the growth pools with a temperature dependant factor f_T, ($0 \leq x_T \leq 1$) depending on the existing extreme temperature (Goudriaan and van Laar, 1994), 3) the frost effect f_r on the reproductive developmental stages are similarly being represented as temperature dependant ($0 \leq f_r \leq 1$) (Drozdov et al., 1984; Single, 1984; Whaley et al., 2004) and is also multiplied by the growth pools. The frost effect factor is 1 (no frost effect), when the daily average temperature for the reproductive phases is higher than or equal to zero, with lower temperature the frost factor decreases till zero (the lowest value). Then the calculations of S_{BM} and RES will be as follows:

$$S_{BM,i} = (S_{BM,i-1} + GR_{SM}) \times \%Y \times x_T \times fr \tag{47}$$

$$RES_i = (RES_{i-1} + R_{RES}) \qquad (48)$$

$$BM = S_{BM} + RES \times \%Y \times x_T \times f_r \qquad (49)$$

The developed external environmental factors (%Y, x_T and f_r) had been added to the growth pools equations of Goudriaan and van Laar (1994).

2.4.4 Partitioning to Crop Organs

The partitioning of the dry matter into the crop organs depends on estimated equations, which have been developed according to the distribution of the timing of the crop yield components (Rawson and Macpherson, 2000), which mentioned in details the timing of the initiation and death of each plant organ depending on the growth stages, which is controlled by the thermal degree units (Figure 9). The second source used for developing the following dry matter allocation equations was a detailed real data set for six experiments showing the dry matter distribution in the plant organs through different thermal degree units (Iglesias, 2006), represented at the Annex 1, 2 and 3. According to Iglasis (2006) data, where the thermal degree units were plotted against the dry matter of the root, leaves, stem and grain, with excluding from it the values that did not match the timing of the plant organs initiation and depth according to Rawson and Macpherson (2000), if exists, however, the Rawson and Macpherson (2000) data were used as a standard performance for the plant organs for adjusting the six detailed experiments of Iglasis (2006). The estimated equations were developed to represent the dry matter allocation to plant organ growth for the six used experiments.

$$R_{DM} = a_1 + \frac{b_1}{1+((RT_{sum}+c_1)/d_1)^2} \times \frac{BM}{100} \qquad (50)$$

$$S_{DM} = a_2 + b_2 \times exp\left(c_2 \left(\frac{RT_{sum}+d_2}{e_2}\right)^2\right) \times \frac{BM}{100} \qquad (51)$$

$$G_{DM} = a_3 + b_3 \times (RT_{sum})^{c_3} \times \frac{BM}{100} \qquad (52)$$

$$L_{DM} = BM - R_{DM} - S_{DM} - G_{DM} \qquad (53)$$

where, R_{DM}, S_{DM}, G_{DM}, and L_{DM} are the root, stem, grain and leaf dry matter [g.dm m^{-2} d^{-1}] respectively.

Table 5 presents the predicted coefficient values for the last parameters for the plant organs. These equations and coefficients have been estimated by using the program SlideWrite Plus for Windows version 3.0, which use both linear and nonlinear algorithms as statistical models to determine the best equation, which fits a set of observations. The r^2 coefficient for the nonlinear regression equations 50 – 52 was 0.99 for all. The equations 50,

51 and 52 for calculating the dry matter allocation of the root, stem and grain, respectively, used the Lorentzian, Gaussian and power nonlinear fit equations, respectively.

The model had been transformed then from the mathematical form to the programmatical form, for developing a windows application that allows the user to use a complex model with a graphical user interface (GUI). The softwares, which have been used for developing the model, were the programming language Microsoft Visual Studio 2010© (Microsoft Basic.NET), which handles all the model equations, conditions and variables, and the database engine Microsoft SQL Server 2008©, which is responsible for organizing all the available data that required by the model in a database, where it will be connected with the model and its data will be available to the model.

Table 5. The plant organs coefficients and their values

Parameter	Coefficient	Value (dimensionless)
R_{DM}	a_1	3.33
	b_1	99.5
	c_1	30.5
	d_1	345.5
S_{DM}	a_2	9×10^{-3}
	b_2	44.16
	c_2	-0.5
	d_2	-1163.5
	e_2	266.86
G_{DM}	a_3	-0.23
	b_3	2.5×10^{-18}
	c_3	6.08

2.5 Model calibration

For the model calibration, the Dürnast – Freising field location and weather station had been used. The field experiments in Dürnast for the studied seasons (2000/01 – 2008/09) used many different cultivars of winter wheat. These field experiments had different water and fertilization treatments. The optimal fertilization treatments and the open field experiments without water treatments had been selected.

2.6 Model validation

After completing the calibration step with a satisfactory result, the simulated yield had been compared with independent dataset, which consists of the yield data from the other eleven experimental locations (Table 2). The model followed also in its simulation the same used planting date, soil type and weather data of the experimental locations. The validation step did not depend only on comparing the simulated yield with different experimental location yield data, but also on different growing seasons (2000/01 – 2008/09) for the same location.

2.7 Model efficiency

The calibration and validation steps of the model needed criteria for evaluating, how successful these steps were, comparing to the real data. For evaluating the efficiency of the model, the Nash-Sutcliffe efficiency (NSE) (Nash and Sutcliffe, 1970) had been used. The NSE compares model predictions to the mean of observed values to determine the better predictor of observed values. Its value is based on the dispersion of variables around the line of equal values. Also the root mean square error (RMSE) – Observations' Standard Deviation Ratio (SR) collectively called RSR was calculated. RSR was developed by Moriasi et al. (2007) based on the recommendation of Singh et al. (2004). This error index criterion is used to quantify error in units of the variable being evaluated. In order to develop a performance rating for RMSE, it was divided by the standard deviation of observed values to create RSR (Sexton, 2007).

$$RMSE = \sqrt{\frac{1}{n}\sum_{i=1}^{n}(o_i - m_i)^2} \qquad (54)$$

$$NSE = 1 - \frac{\sum_{i=1}^{n}(o_i-m_i)^2}{\sum_{i=1}^{n}(o_i-\bar{o})^2} \qquad (55)$$

$$RSR = \frac{RMSE}{STDEV_{O_i}} = \frac{\sum_{i=1}^{n}(o_i-m_i)^2}{\sum_{i=1}^{n}(o_i-\bar{o})^2} \qquad (56)$$

where, n is the number of the studied years at each or all locations, o_i is the ith real value being collected, m_i is the ith simulated value for that being evaluated, $\bar{o}$ is the mean values and σ_0 standard deviation of the real data.

The value of NSE ranges from negative infinity (poor model) to 1.0 (perfect model). If NSE<0, the observed mean is a better predictor than the model; NSE=0, the observed mean is as good a predictor as the model; NSE>0, the model is a better predictor of observed data than the observed mean (Legates and McCabe, 1999; Wilcox et al., 1990). According to Moriasi et al. (2007), very good to satisfactory values of NSE fall in the range of 1 to 0.5.

The resulting criterion and expected values of RSR can then be applied to various constituents. The value of RSR ranges from 0 (perfect model) to a large positive value (poor model). According to Moriasi et al. (2007), very good to satisfactory values of RSR fall in the range of 0.0 to 0.7.

2.8 Future weather data

The model had been also built for simulating the expected wheat yield under the predicted future weather data. The predicted future weather data were collected from the world data center for climate (WDCC). The World Data Center (WDC) system was created to archive and distribute data collected from the observational programs of the 1957-1958 International Geophysical Year. Originally established in the United States, Europe, Russia, and Japan, the WDC system has since expanded to other countries and to new scientific disciplines. The WDC system now includes 52 Centers in 12 countries. Its holdings include a wide range of solar, geophysical, environmental, and human dimensions data. The WDCC is maintained by Model and Data (M&D). The WDCC is included in the CERA database system. CERA (Climate and Environmental Retrieving and Archiving) is a database model that has been designed by PIK (Climate Impact Research Institute in Postdam), DKRZ (Climate Computing Centre), and AWI (Alfred Wegener Institute, Bremenhaven) to enable interchange of meta information on geo-referenced data (WDCC, 2008).

Three models have been used from the CERA database, REMO, CLM, and STARII. REMO and CLM are regional climate models, which simulated future weather data each 0.088 and 0.165 degree horizontal grid resolution respectively, for the latitude and longitude. Both models had available predicted data from 2001 to 2100. STARII model is a statistical regional model, which had available predicted data for the period 2007 – 2060. The used scenarios for the last three models were A1B, which describes a possible future world of very rapid economic growth, global population peaking in mid-century and rapid introduction of new and more efficient technologies with a balance across all energy sources.

At the predicted future data, the last twelve studied locations for the crop model in Table 1 have been also used for the comparison between the current and future weather conditions on the winter wheat. For the regional models (REMO and CLM), the twelve studied sites were collecting by the average of 25 predicted sites that located around the required one, where the studied site is in the middle of the intercept of five consequences longitude and latitudes. This interception gave 25 point with 100 years future weather data for each point. The averages of the weather elements of these points are representing the studied site. The required weather elements are the maximum and minimum temperature, relative humidity, precipitation, and solar radiation. For the STARII model, five locations that are surrounding the studied one have been chosen to represent it, taking into consideration that these five locations are not to be far from the studied one. The averages of these locations' elements have been also calculated to represent the real one in the future. The predicted future weather data effect on the winter wheat had been divided into three periods; 2021 – 2050, 2051 – 2080, and 2071 – 2100. These three periods are applied on the REMO and CLM model, but for the STARII model, only the first period is applied, because of the availability of the STARII future weather data, which is only till 2060. There is an overlap between the second studied period (2051 – 2080) and the third (2071 – 2100) to use all the available weather data for studying its effect on the winter wheat.

The used weather elements at the three future models are: the maximum and minimum temperature [°C], precipitation [mm], solar radiation [MJ m^{-2} day^{-1}], relative humidity [%], and wind speed [m.s^{-1}]. The relative humidity element (RH) at the CLM is absent; therefore it has been calculated with the following equations according to (Kümmel, 1997):

$$RH = e/es \times (100\%) \tag{57}$$

$$e = es_o \times exp\left(\frac{lv}{Rv} \times \left(\frac{1}{T_o} - \frac{1}{T_d}\right)\right) \tag{58}$$

$$es = es_o \times exp\left(\frac{lv}{Rv} \times \left(\frac{1}{T_o} - \frac{1}{T_k}\right)\right) \tag{59}$$

where, e is the environmental vapor pressure [hPa], es is the saturation vapor pressure [hPa], es$_o$ is the reference saturation vapor pressure (es at a certain temp, usually 0 degree C) = 6.11 [hPa], T$_o$ is the reference temperature (273.15 Kelvin), T$_d$ is the dew point temperature [Kelvin], T$_k$ is the air temperature [Kelvin], lv is the latent heat of vaporization of water (2.5 * 10^6 J. kg^{-1}), and Rv is the gas constant for water vapor (461.5 J.K.kg^{-1}).

2.9   The expected future geographical distribution of yield

An extra window from the developed model was created for simulating the grain yield of winter wheat for all Bavaria by using the same regional models REMO and CLM for each 0.1

and 0.2 degree horizontal grid resolution, respectively for the latitude and longitude. For calculating the expected yield in all Bavaria for the last studied future periods (2021 – 2050, 2051 – 2080, and 2071 - 2100), the simulated yield was calculated from 1) the regional models REMO and CLM by using the A1B scenario, 2) the soil attributes for all Bavaria, which were collected and calculated by the LfU (Bavarian Environment Agency) and the European Soil Data Center (ESDAC, 2012), and 3) the altitude data for the used latitude and longitude in Bavaria was collected from the Bavarian Surveying Administration (Bayerische Vermessungsverwaltung).

The Bavarian Land Office for the Environment (LfU) develops concepts for sustainable and eco-friendly ground use by researching, monitoring and mapping Bavarian soils. These concepts are based on the Bavarian Ground Conservation Program, taking the various interests of economics, politics and science into account. The collected Bavarian soil types by the LfU were described as words (Silt Loam, loamy Sand, etc…), which have been translated to the sand, clay, and silt percentage and the bulk density of each soil type depending on Pälchen (1996) and Renger et al. (2008). The soil types by the LfU were represented in a map and the data were collected by using the ArcGIS© 10.0 as single values depending on the latitude and longitude. The content of organic carbon in the soil collected by the ESDAC was represented in a map and the data was collected by using the ArcGIS© 10.0 as single values depending on the latitude and longitude.

After collecting the weather, soil and location data for the two regional models used in the common data for the last three resources, according to the longitude and latitude, there were 845 points for REMO and 212 for CLM in Bavaria. The collected future weather data for the models REMO and CLM, the soil attributes and the location data for Bavaria were formatted then to be entered in a database, from which the developed model can drag its selected data for calculating the predicted yield for the selected future period across Bavaria. The values of the estimated yield for the CLM and REMO models for the three studied future periods are then translated to a map for showing the expected geographical winter wheat yield distribution in Bavaria by using the ArcGIS© 10.0 program. The expected appearance of the extreme temperature and the grain-filling duration for the studied future period were also calculated and geographically distributed in Bavaria by using the same program.

3 Results

The new crop model was developed to estimate the grain yield of winter wheat in Bavaria, where this estimation depends on the weather conditions (temperature, relative humidity, wind speed, radiation and rainfall) and soil attributes (sand, clay, silt, and organic carbon content in the soil), which were the key input data to simulate the variation of yield variability during the different seasons and different locations.

The model has simulated winter wheat yield, by using the same conditions, which existed during the experiments of the LfL such as the same weather data of the same year, the same soil characteristics and the same planting date. Therefore, the comparison between the simulated and observed yield will show the accuracy of the model. Nevertheless, there were also missing data for comparison. The bulk density and the electric of conductivity of the soil were not available and were estimated with 1.5 g/cm3 for the soil density and 2 dS/m for the soil electric of conductivity. Also the different cultivars effect is not yet taken into consideration.

3.1 Model efficiency

The comparison between the simulated and observed yield is shown first by the model efficiency calculation by using the NSE, which had the value 0.31 across all studied and simulated seasons for all the studied locations together. The NSE value, which lies between 0 and 1, assumed to have a good predictor model to the observed data than the observed mean, and therefore, the model is generally viewed to have a satisfactory acceptable level of performance, but still the value is relatively low within the acceptable range. The total RSR value for all the seasons and locations had the value 0.83, which is a little bit higher than the very good to satisfactory performance range (0.7), where the lower values here are more efficient than the higher ones. For the efficiency details for each studied location, Table 6 shows the NSE and RSR for each site. As clarified in Table 6, the highest value of NSE and RSR were at Landshut and Donau-Ries with values of 0.80 and 0.79 for NSE, respectively, which were close to the perfect match 1, and with values of 0.44 and 0.45 for RSR, respectively, which were in the very good performance range. The Donau-Ries chart (b) at Figure 1 showed also how close the simulated yield is compared to the observed one. The simulated yield at Donau-Ries had values, which were exactly within the observed yield range of the studied cultivars, except for one season (2005/06), where the simulated yield was higher than the observed one with a range from 10 to 20 dt/ha and a slight increase of the estimated yield with 3.5 dt/ha compared to the observed yield at the season 2008/09. This highest NSE value for Donau-Ries had been calculated for the seasons from 2002/03 to 2008/09, because of the lack of the weather data before this period. Also the Landshut (h) simulated yield showed an exact to a very close match to the observed yield range for all the seasons from 2000/01 to 2008/09, except for one season 2006/07, which showed an observed yield increase compared to the previous season 2005/06 with almost 15 to 20 dt/ha, although the simulated yield had almost the same yield value at these two seasons. The NSE for the sites Main-Spessart, Regensburg, Würzburg, and Passau had also acceptable values, which fluctuated in the middle of the range from 0 to 1, with values of 0.54, 0.51, 0.38, and 0.36, respectively, and values of 0.68, 0.70, 0.79, and 0.80, for the calculated RSR, respectively. The NSE and RSR values for the sites Main-Spessart and Regensburg were still in the satisfactory performance range according to Moriasi et al. (2007), but the values for the sites Würzburg and Passau were higher than the RSR satisfactory performance range with a differences of 0.09 and 0.10, respectively, and lower than the NSE satisfactory performance range with differences of 0.12 and 0.14, respectively. At Regensburg, the simulated yield behavior was close to the observed one during the studied seasons with no or slight deviation from the observed yield range. Two seasons showed a different behavior from the observed yield range, the first was at the season 2005/06, where the observed yield range had slightly increased compared to the previous season 2004/05, but the simulated yield decreased, and showed a difference between the observed and simulated one with 11-17 dt/ha. The second different season was 2008/09, which showed an increase of the simulated yield compared to the previous season, while the observed yield decreased, resulting in a simulated yield that was higher by 8-22 dt/ha compared to the observed one. This overestimation of the simulated yield during the season 2008/09 occurred also at the sites Günzburg, Passau, Würzburg, Weißenburg-Gunzenhausen, and Neumarkt differing by 8-13, 16-22, 5, 20, and 13-25 dt/ha, respectively. The simulated yields at Main-Spessart and Eichstätt for all the studied seasons were similar to the observed yield with no or only slight deviations of 1-3 dt/ha, but with significant differences such as 14-18 and 6-9 dt/ha at the seasons 2004/05 and 2008/09 in Main-Spessart, respectively, and 8-24 and 11-29 dt/ha at the seasons 2002/03 and 2003/04 in Eichstätt, respectively. Nevertheless, the NSE of Würzburg was relatively low, the graph for Würzburg displayed the same performance in yield between the modeled and observed one for all the studied seasons with no or small differences, except for the season 2007/08, where the observed yields were higher than the simulated by 9-15 dt/ha. Hence, the NSE decreased in its value by 0.30, 0.28, 0.26 and 0.26, and the RSR increased in its value by 0.84, 0.85, 0.86 and 0.86 at the sites Weißenburg-Gunzenhausen, Freising, Günzburg, and Eichstätt, respectively. These NSEs' values were still within the range of the acceptable performance, but with quite small values, and also had higher values of RSR than the satisfactory performance range, although the modeled yield behavior at these sites showed a quite good performance with the observed yield. A close behavior of yield between the modeled and observed was also observed for Weißenburg-Gunzenhausen for the seasons from 2000/01 to 2005/06, but the season 2006/07 showed a different performance with differences of 9-16 dt/ha, together with the 2008/09 season, as mentioned before. At Freising, only the two seasons 2001/02 and 2006/07 showed a different behavior between the simulated and observed yield, with differences by 9-23 and 10-22 dt/ha, respectively. The season 2001/02 indicated a significant difference between the modeled and observed yield with values of 11-26 dt/ha at Günzburg and at Passau for the seasons 2001/02, 2002/03 and 2006/07 with values of 7-22, 6-24, and 15-19 dt/ha, respectively, plus the 2008/09 season, as mentioned before. The last two sites Lichtenfels and Neumarkt, which showed almost a zero NSE value and a RSR value of one, and indicated significant differences between the modeled and the observed yield with the same yield behavior with values of 13-26, 9-31, and 9-15 dt/ha at the seasons 2002/03, 2003/04 and 2008/09, respectively, for Lichtenfels, and 9-21 and 7-15 dt/ha at the seasons 2002/03 and 2006/07, respectively, for Neumarkt, and with the opposite behavior with values of 8-26 and 29-39 dt/ha at the seasons 2005/06 and 2006/07, respectively, for Lichtenfels, and 24-30, 17, and 13-25 at the seasons 2004/05, 2007/08, and 2008/09, respectively, for Neumarkt.

Confidence ranges between the simulated and the average observed yields for all studied cultivars are given in Figure 2, aiming to clarify how close or far the simulated yield to the observed one was. Figure 2 shows that the simulated yield values (symbols) were within the +20% and -20% range of the average observed yield values. The distribution of the simulated yield values were sometimes exactly on or close to the average yield line, like at Landshut and Donau-Ries except at the seasons 2006/07, and 2005/06, respectively, but in general, the distribution of the simulated yields were located between the +20% and -20% lines. In Eichstätt, the simulated yields were all distributed between the average yield line and the -20% line, which means that all the simulated yields at that site were underestimated compared to the observed yields, with a range from 0 to 20 dt/ha. On the contrary, at Main-Spessart the simulated yields in general were overestimated compared to the observed yield, where all the simulated yields were spreading between the average yield line and the +20% with values ranging from 0 to 15 dt/ha far from the average yield line, except for the season 2005/06, which was underestimated by 15 dt/ha. The simulated yields that were beyond the +20% and -20% lines were at the sites Lichtenfels and Neumarkt, leading to almost zero NSE values. At the Lichtenfels site results of four simulated seasons 2002/03, 2004/05, 2007/08, and 2008/09 lay nearby outside the range and the yields were all overestimated. At Neumarkt yield of the two seasons 2005/06 and 2006/07 were underestimated and lay also outside the (-20%, +20%) range. There were three other overestimations of yields outside the (-20%, +20%) range, for the seasons 2002/03 and 2008/09 at Passau, and 2008/09 at Weißenburg-Gunzhausen.

Table 6: The Nash-Sutcliffe efficiency (NSE) and the root mean square error (RMSE) for the Observations' Standard Deviation Ratio (RSR) values for the studied locations.

Location	NSE	RSR	Location	NSE	RSR
Würzburg	0.38	0.79	Donau-Ries	0.79	0.45
Passau	0.36	0.80	Regensburg	0.51	0.70
Lichtenfels	-0.10	1.05	Eichstätt	0.26	0.86
Landshut	0.80	0.44	Neumarkt	0.06	0.97
Weißenburg-Gunzhausen	0.30	0.84	Main-Spessart	0.54	0.68
Günzburg	0.22	0.86	Freising	0.28	0.85

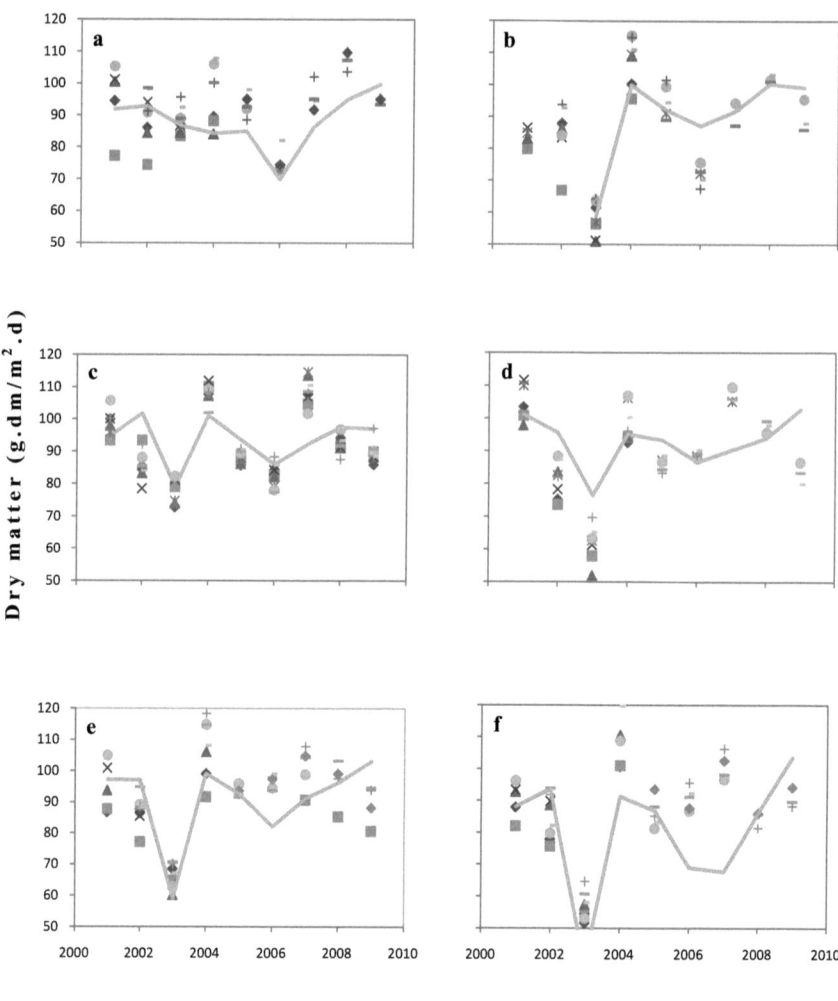

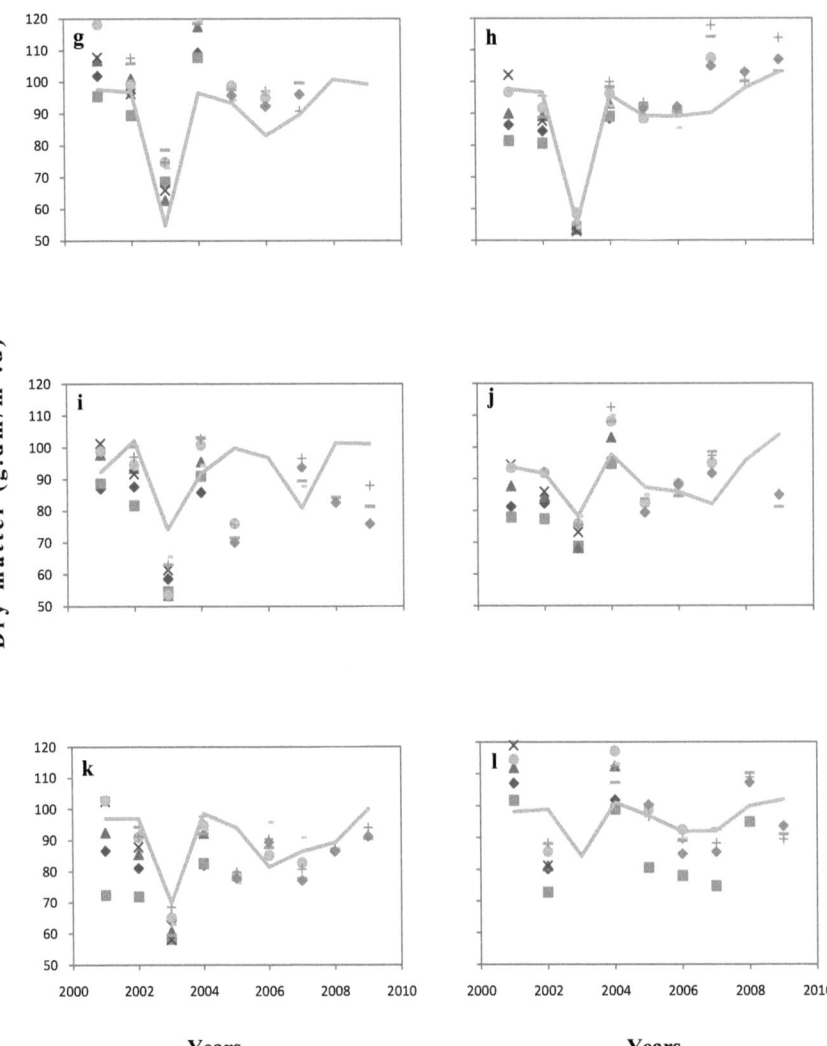

Figure 1. Comparison of simulated (——) and measured values (the symbols) for grain yield of different wheat cultivars grown at Würzburg (a), Donau-Ries (b), Freising (c), Passau (d), Regensburg (e), Lichtenfels (f), Eichstätt (g), Landshut (h), Neumarkt (i), Weissburg-Gunzenhausen (j), Main-Spessart (k), and Günzburg (l) during the seasons from 1999/2000 to 2008/09.

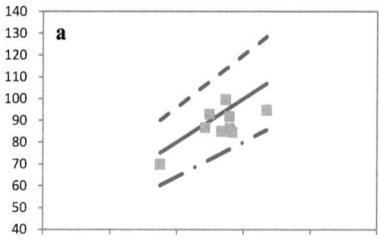
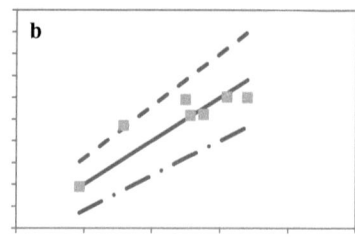
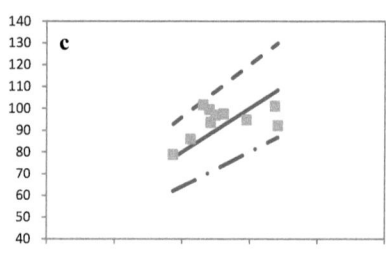
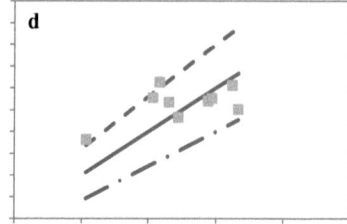
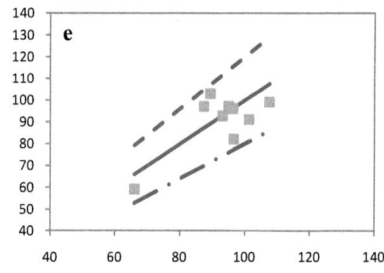
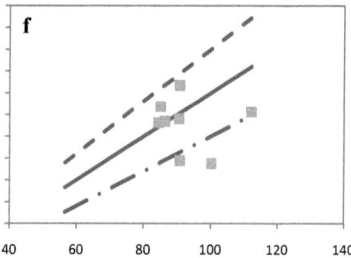

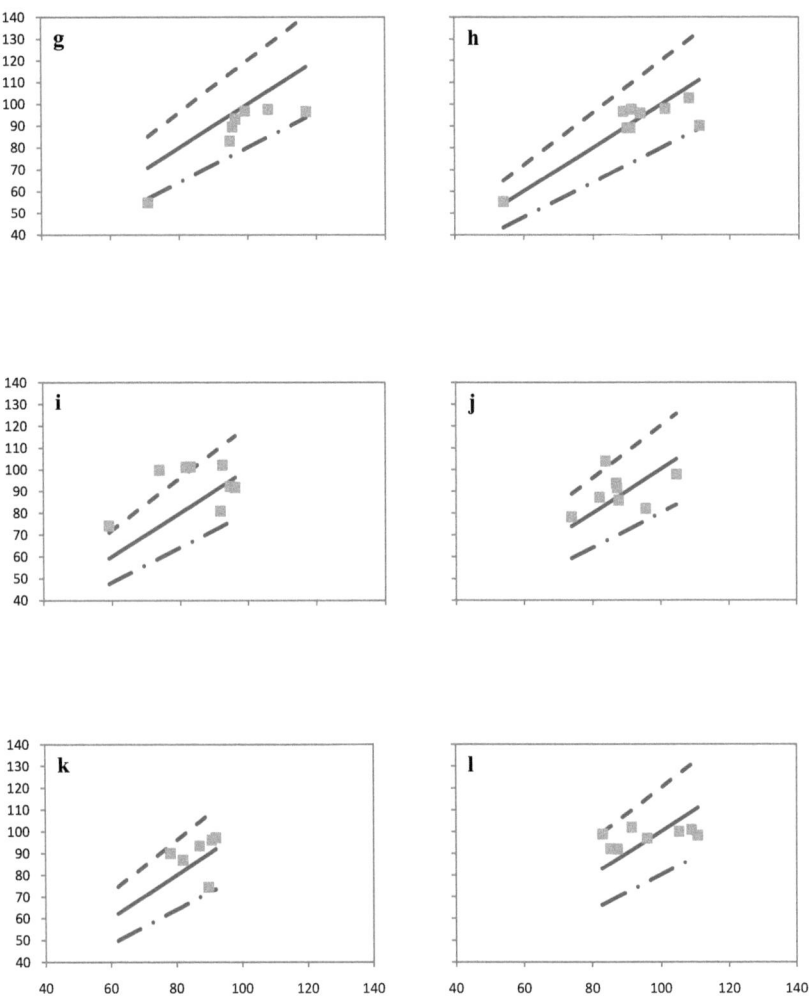

Figure 2. Comparison of simulated (■), measured (——), +20% of the measured values (- - ·) and -20% of the measured values (— ·) for the grain yield average of the used cultivars vs. the measured average grain yield at Würzburg (a), Donau-Ries (b), Freising (c), Passau (d), Regensburg (e), Lichtenfels (f), Eichstätt (g) , Landshut (h), Neumarkt (i), Weissburg-Gunzenhausen (j), Main-Spessart (k), and Günzburg (l) during the seasons from 2000/01 to 2008/09.

3.2 The simulated biomass partitioning

The simulated partitioning of the winter wheat crop biomass is shown in Figure 3. The simulated partitioning indicated the performance of the aboveground biomass and its allocation to the grain and straw (stem and leaves). Landshut was used as an example for displaying its simulated partitioning rules affected by different weather conditions during the studied seasons from 2000/01 to 2008/09. The accumulation of the thermal degree units and hence of the dry weight accumulation timing of the different plant organs varied depending on the existing weather data, including the number of the heat waves and water deficit periods, which faced the planting season. The simple partitioning of the other studied sites were shown as appendix information (A1). The soil type of the studied site Feistenaich in Landshut is silt loam, which is a very good soil, therefore, the water deficit periods were not intensive at the site during the nine studied seasons. Landshut had been chosen, where it had the best NSE (0.80) almost like Donau-Ries (0.79). Also both sites Landshut and Donau-Ries had the lowest RSR values, which were in the very good performance range. But because of the lack of weather data for Donau-Ries till the end of 2001, the yield was estimated starting from the season 2002/03. Therefore, Landshut was a better choice to be represented, according to the variability of the weather, soil and plant data during the studied seasons.

The first two seasons 2000/01 and 2001/02, as shown in Figure 3 had almost the same weather effect on the plant, both faced two heat waves. The first heat wave was only one day at the season 2000/01 at 31/7 with a temperature of 31.6°C and two days for the season 2001/02 at 19 and 20/6 with temperatures of 33 and 32°C, respectively. The second heat wave was two days for the first season at 15 and 16/8/2001 with temperatures of 32 and 32.3°C, respectively, and one day at 2001/02 season at 9/7/2002 with a temperature of 32.5°C. The period between the sowing stage and emergence stage was 10 days at both seasons, and the leaves initiation started one week before November, probably due to the optimal temperature range at that stage, which differed by 5 to 13°C at the 2000/01 season and by 4 to 13°C at the 2001/02 season, and no water deficit was apparent at the two seasons. The lowest yield was observed at the season 2002/03, as a result of the water deficit and several heat waves appeared during the season. Seven heat waves with 11 hot days were observed during that season at 6/5 with 31.1°C, 5/6 with 31.7°C, 10 and 12/6 with 32.2 and 34°C, respectively, 23/6 with 36.4°C, 29 and 30/6 with 32.1 and 34.5°C, respectively, 16/7 with 33.4°C, and at 20, 21, and 22/7 with maximum temperatures of 33.4, 33.3, and 33.2°C, respectively. Additionally the growth had been also affected by a strong water deficit period, which appeared on the date 12/6, while that period suffered also from a heat wave. The occurrence of many heat waves and water deficit periods and their impact on the crop behavior by decreasing the biomass noticeably at mid of June, where the water deficit and 2 heat waves appeared and less marked on the other heat waves dates. Also after the sowing date (5/10/2002) until the emergence stage, which was only a week, the average temperatures were relatively warm and the leaves dry weight reached almost 50 g.dm/m2 at the first week of November, and with regard to the warm weather the leaves' growth increased sharply after the middle of April before facing seven heat waves, therefore the dry weight decreased. The season 2003/04 was a high yield season, which did not suffer from any water deficit, and only 2 late hot days at 20/7 and 12/8 with maximum temperatures of 31.2 and 31.9°C, respectively, which had slight negative effects on growth. The season 2003/04 was a cold season, although the sowing date was at the end of October with a long sowing stage period of 20 days, causing the leaves dry weight during the leaves initiation to increase only little before reaching the dormancy period during the winter. The accumulation of the dry matter in general during this season till June occurred relatively slowly, depending on the daily temperature. The season 2004/05 had been affected by comparatively the same temperature effect as the previous season 2003/04, starting from the late sowing date (30/10), a long sowing stage till the emergence stage (14 days), a reduced leaves initiation, no water deficit appearing, and the relatively slow accumulation of the dry matter related to the relatively low season temperature, but starting from the end of May the season faced four heat waves, which decreased the total biomass and also the final yield by almost 7 dt/ha compared to the last season. These heat waves were relatively pronounced with a noticeable negative effect on the total biomass and yield. The heat waves started at 28 and 29/5 with maximum temperatures of 36.3 and 36.8°C, respectively, 27, 28 and 29/7 with temperatures of 31.2, 33.9, and 34°C, respectively, and 12/8 with 32.5°C and 19/8 with 35.4°C. Two long late successive heat waves appeared at the seasons 2005/06 and 2006/07, for the first season the heat waves were at 20, 21 and 22/7 with maximum temperatures of 32.7, 31.9 and 31.8°C, respectively, and at 25, 26, and 27/7 with 31.9, 32.2 and 32.5°C, respectively, and for the second season at 15 and 16/7 with 32.2 and 34.7°C, respectively, and at 19 and 20/7 with maximum temperatures of 31.4 and 31.2°C, respectively. These two seasons had similar sowing stage durations, 8 days at 2005/06 and 9 days at 2006/07, respectively, because of the relatively warm temperature at the sowing stage (6-13°C), although the season 2006/07 showed a warmer temperature at the

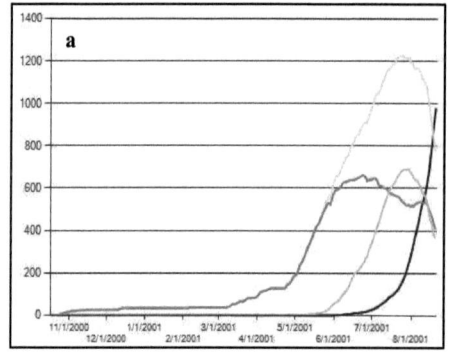
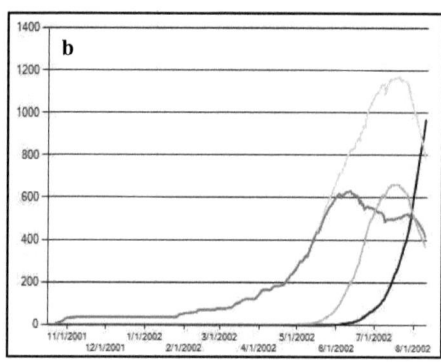
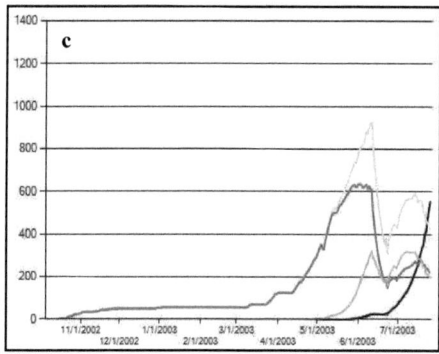
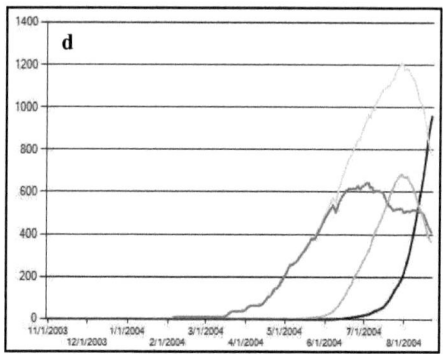
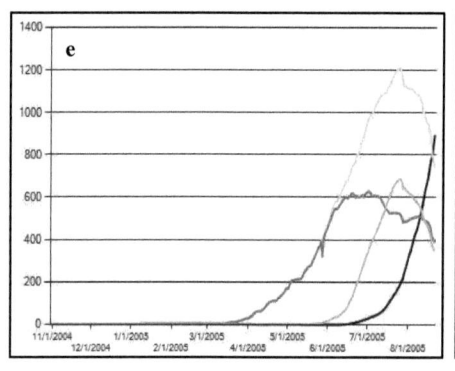
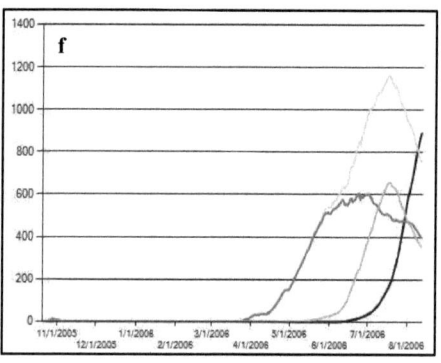

Date Date

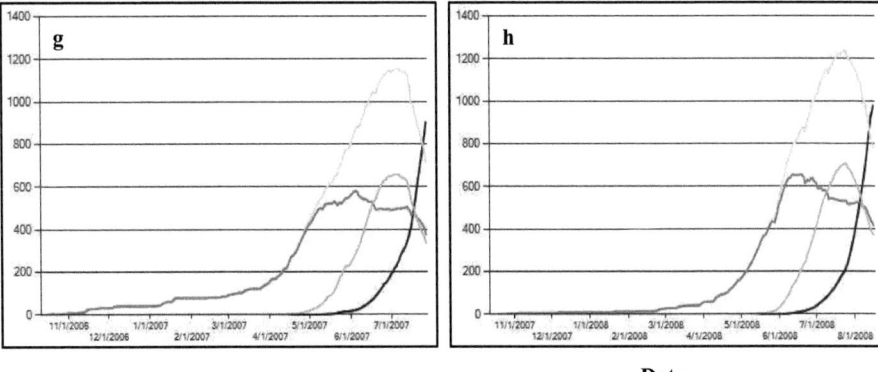

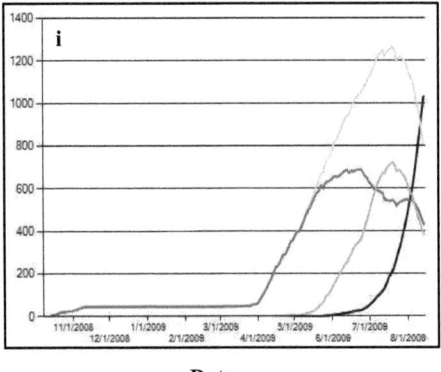

Figure 3. Simulated course of biomass partitioning to leaves (——), stem (——) as well to grain (——) and straw (——) yield (Dry matter (g.dm/m^2.d)) at Landshut in the seasons 2000/01 (a), 2001/02 (b), 2002/03 (c), 2003/04 (d), 2004/05 (e), 2005/06 (f), 2006/07 (g), 2007/08 (h), and 2008/09 (i).

emergence stage, while the season 2005/06 was characterized by a cold one. Therefore, the leaves initiation of the season 2006/07 started earlier than the previous season. An early water deficit period appeared at the season 2006/07 at 21/10, which did not apparently affect growth. Although the season 2007/08 had relatively cold and long sowing (20 days) and emergence stages, which resulted in a small leaves dry matter before the dormancy period, but in a relatively long leaves initiation stage, and slow dry matter accumulation till the first of May. But also the season faced three hot days, which were at 11/7, 31/7, and 7/8 with maximum temperatures of 31.1, 31.1, and 32.3°C, respectively. These hot days affected slightly the final yield, while this season led a high yield with 98 dt/ha. With optimal growth temperatures, a relatively long dormancy period, no water deficit, and only one hot day at 23/7 with maximum temperature of 34.2°C, the season 2008/09 was characterized by the highest yield compared to the other studied seasons with a value of 103 dt/ha.

With the same concept, the dry matter allocation behavior of the other studied sites was affected by different environmental conditions (soil characteristics and weather data). Landshut is characterized by a good soil type and suitable weather conditions for the growth, therefore, the yields in Landshut were high during the studied seasons except for 2002/03, which was the only season that faced a water deficit and many hot episodes during plant growth. At the other studied sites (Appendix, A1), the appearance of hot episodes and water deficit cases were higher than in Landshut, and they appeared severely in the dry season 2002/03. At Donau-Ries the season 2002/03 faced four following water deficits periods from 29/3/2003 to 8/5/2003, plus 17 hot days starting directly before the anthesis at 12/6 till the end of the grain-filling stage. All these factors decreased the yield at that season to 58.9 dt/ha. Also in Lichtenfels at the same season, which faced five water deficits periods during almost the whole developmental stages and 14 hot days from the terminal spikelet stage till the grain-filling stage, resulted a very low yield of 38.3 dt/ha. With the same consequence at the same season, Eichstätt revealed a total yield of 54.8 dt/ha after facing four water deficits at 29/3, 18/4, 28/4, and 7/6 and eight hot episodes starting from 31.5 to 34.7°C. Freising also faced ten hot days with two water deficits and obtaining a final total yield of 79 dt/ha at the season 2002/03. At Neumarkt that season had four water deficit periods, which occurred all before the double-ridge stage, and 15 hot days with a total yield of 74 dt/ha. Passau, Regensburg, and Weissenburg-Gunzenhausen faced all two water deficit periods at the season 2002/03, and further 11, 19, and 11 hot days, respectively, with total yields of 76.3, 59.2, and 78 dt/ha, respectively, while Main-Spessart faced three water deficit periods plus

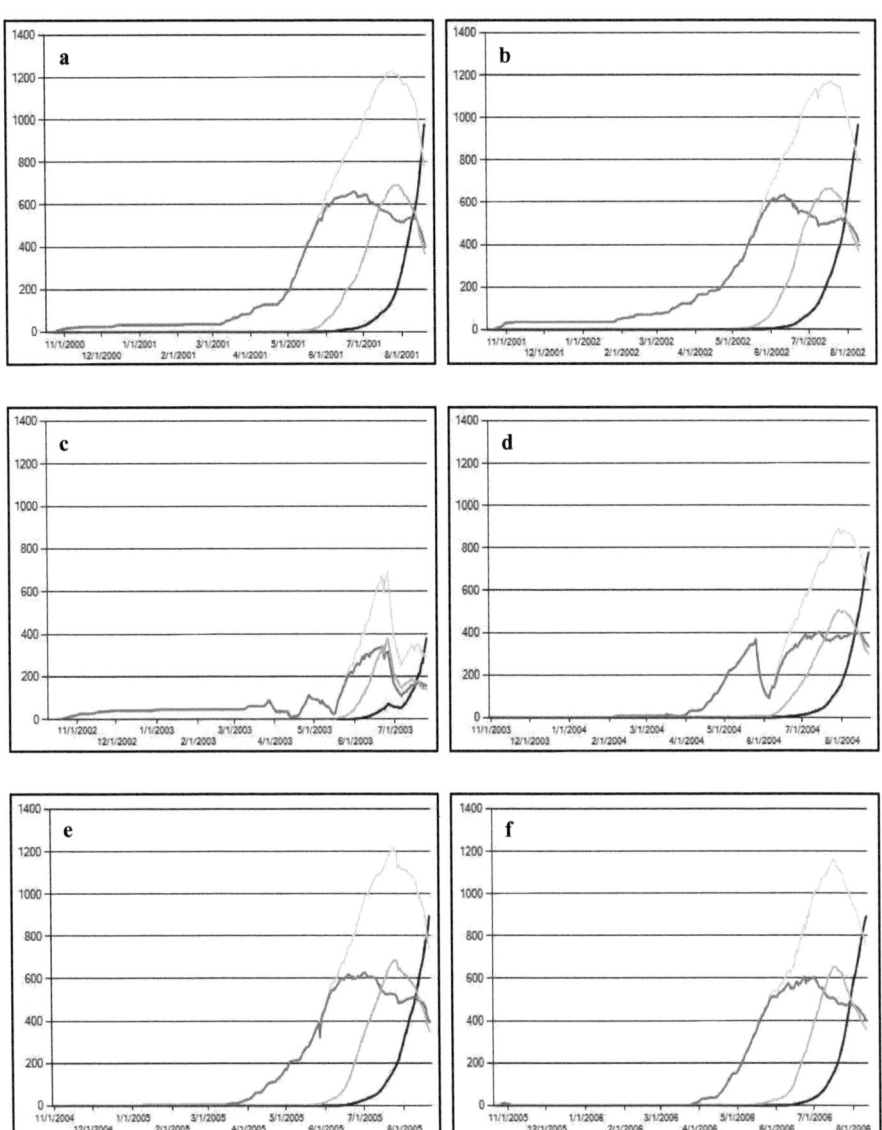

Date

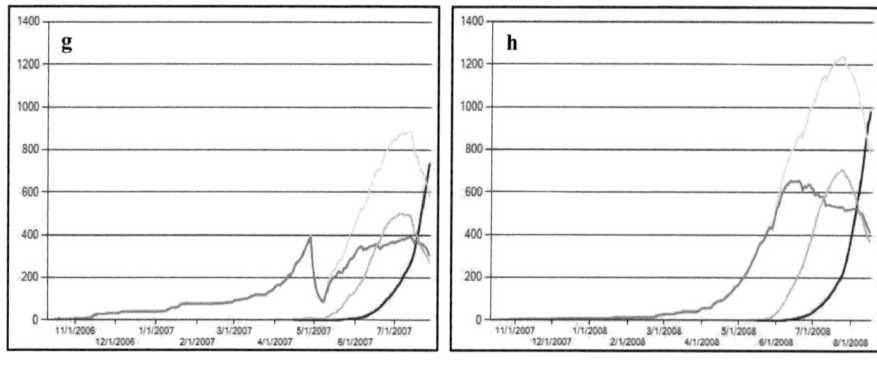

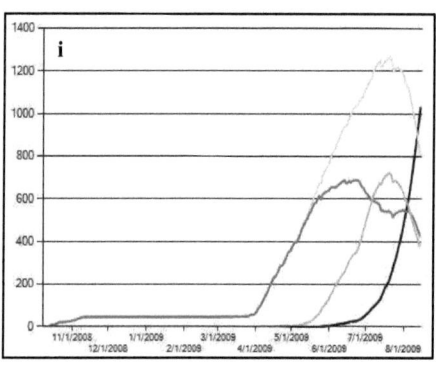

Date

Figure 4. Simulated course of biomass partitioning to leaves (——), stem (——) as well to grain (——) and straw (——) yield (Dry matter (g.dm/m^2.d)) at Landshut in the seasons 2000/01 (a), 2001/02 (b), 2002/03 (c), 2003/04 (d), 2004/05 (e), 2005/06 (f), 2006/07 (g), 2007/08 (h), and 2008/09 (i), by using a sandy loam soil.

twelve hot days and had 69.9 dt/ha total yield. Not only the season 2002/03, which faced water deficit periods with hot episodes, but in Lichtenfels, the seasons 2005/06, and 2006/07 faced two water deficit periods, and fourteen hot days, and one water deficit period and four hot days, respectively, with total yields of 68.8 and 67.3 dt/ha, respectively. Passau faced seventeen hot days during the season 2005/06 starting from a week before anthesis till the grain-filling stage and its total yield was 86.5 dt/ha. As well as, Main-Spessart had twelve hot days with no water deficit period, and eight hot days with one water deficit period at the season 2005/06 and 2007/08, respectively, and the total yield at these seasons were 81.1 and 89 dt/ha, respectively, while in Würzburg the season 2005/06 had a low yield of 69.8 dt/ha after facing during the growing period fifteen hot days and a water deficit period. At Weissenburg-Gunzenhausen, very high temperatures characterized the seasons 2004/05 and 2006/07 with temperatures of 36 and 38.3°C, respectively, where the total dry matter declined after these hot waves with values to 6 and 14 dt/ha, respectively.

The water deficit periods showed up in some cases at the beginning of the growing season, at the sowing stage and the beginning of the emergence stage, and these water deficit periods did not affect significantly on the yield. The appearance of the early water deficit periods appeared mainly at the seasons 2005/06 and 2006/07. At the first season (2005/06), early water deficit periods existed in Donau-Ries at 24/10, in Eichstätt at 24/10, in Freising at 21 and 24/10, in Lichtenfels at 23/10 and 2/11, in Regensburg at 21/10, in Main-Spessart at 27/40, which was the beginning of the emergence phenological stage date, and in Weissenburg-Gunzenhausen at 27/10, and in Würzburg at 20/10. The same last sites faced also an earlier water deficit period, but at the season 2006/07 except for Lichtenfels and Würzburg, which faced no earlier water deficit periods, but it appeared in Günzburg at the season 2006/07 at 22/10, which was the beginning of the emergence phenological stage date, nevertheless this location did not face an earlier water deficit period at the season 2005/06.

The timing of the developmental stages differed depending on the weather data. Therefore, the differences between the beginning of the sowing and the emergence stages during the different seasons and sites were from one to two weeks, while it lasted for a month to 40 days at some cases like in Donau-Ries, Günzburg, Lichtenfels, Passau, Main-Spessart, and Weissenburg-Gunzenhausen at the season 2003/04, but in other cases the period between the sowing and emergence stages was very long as in Neumarkt for the seasons 2002/03, 2003/04, and 2007/08 with values of 123, 111, and 95 days, respectively, and in Freising and Regensburg at the season 2003/04 with values of 91 and 113 days. The beginning of the double-ridge stage was normaly in April till the beginning of May, but it was earlier in January and February in some cases, at the season 2006/07 in Donau-Ries, Lichtenfels, Main-Spessart, Weissenburg-Gunzhausen, and Würzburg at the dates 24/1, 18/2, 19/1, 19/1, and 10/1, respectively, and at the season 2000/01 in Main-Spessart at 13/1. The harvest date was late at the season 2003/04 for all the studied locations, where it fluctuated from 16/8 at Würzburg to 27/9 at Neumarkt.

3.3 Soil water content sensitivity analysis

The crop performance is affected by the different weather data and soil attributes. The fluctuation of the total biomass and the yield at Landshut during the studied nine seasons in Figure 3, was affected mainly by only the weather data, where the soil type is the same at the same studied site. For analyzing only the effect of the soil attributes on the total biomass and yield, the weather data should be also fixed, therefore, the studied site Landshut had been run again by the crop model using the same studied nine seasons, but with changing the used soil characteristics. The soil type of Landshut was mentioned in LfL data as silt loam soil, which was used as in Figure 3, while the soil type changed to be sandy loam at the same studied seasons. Figure 4 displayed the performance of the aboveground biomass and its allocation to the grain and straw (stem and leaves) of Landshut under the studied nine seasons by using the sandy loam soil. As mentioned before, the yield at Landshut site during the nine studied seasons were high, except at the season 2002/03, with no appearance of water deficit periods, also except for the hot season 2002/03, according to the very good soil type at this site, in addition to the relatively favorable weather at many of the studied seasons. The crop performance was the same at the two soil types in Figures 3 and 4 at many of the studied seasons 2000/01, 2001/02, 2004/05, 2005/06, 2007/08, and 2008/09, where these seasons did not suffer severely from low precipitation and also they faced a very few hot days during all seasons fluctuating from 1 to 3 hot days, except at the seasons 2004/05 and 2005/06, they faced 7 and 6 hot days, respectively. The other seasons 2002/03, 2003/04 and 2006/07 showed water deficit periods with the sandy loam soil, which is not as good as compared to the silt loam. The last seasons gave also lower yields compared to the same seasons with the silt loam soil with almost 20 dt/ha for all. The season 2002/03 revealed five water deficit periods instead of one with the new soil type, these periods ended at 30/3, 9/4, 29/4, 9/5, and 28/6, where the precipitation amounts during the months from February to June were 10.2, 17.7, 39.5, 60.6, and 29.1 mm. Four water deficit periods faced the season 2003/04, which decreased also its yield with nearly 20 dt/ha compared to using the silt loam soil. The first two water deficit periods appeared early at the sowing stage, where they ended at 9/11 and 19/11, and for the other two the water deficit periods were in the double-ridge and terminal spikelet stages at 18/3 and 27/5, respectively. The precipitation amounts at this season were relatively low for the months November, February, and April with values of 27.9, 18.3, and 25.3 mm, respectively. The last different season was 2006/07, which faced 2 water deficit periods; the first one was early in the sowing stage at 21/10, and the second at 29/4 in the terminal spikelet stage, which was a severe one and indicated a significant decrease in the aboveground biomass. The second water deficit period appeared when the precipitation amount was relatively low at March with values of 29.5 mm, followed by a dry month (April), which showed only 6 mm amount of precipitation.

3.4 Scenario simulation based on climate projection for 2021-2100

The model had been also run using projected weather data. These weather data was taken from the CERA database for the models REMO, CLM as regional models and STARII as a statistical model. The model was applied to weather data for 3 periods: (1) 2021 – 2050, (2) 2051 – 2080, and (3) 2071 – 2100 for REMO and CLM, and only the first weather period for STARII, according to the lack of the data in STARII. The simulated average yields under the three projected weather periods are shown in Figure 5, and individually in Figure 6. In general, by using the STARII weather data, the developed model here showed always the highest simulated yields than for the models REMO and CLM for the first predicted period (2021 – 2050), but at the other two future periods (2051 – 2080 and 2071 – 2100) there were no data available for the STARII data to be applied, therefore the comparison was only between the three used models during the first predicted future period. The crop model showed the highest yield by using STARII data with significant values compared to the data of the other two models, except for Lichtenfels, where the simulated average yield at the first period under the STARII model was higher than under the REMO data average yield at the same period with only differing by 2.4 dt/ha, and was higher than for the CLM model with an average yield of 7.3 dt/ha. At the other sites, under the CLM data the average yield for the first predicted period was higher compared to the average yield for REMO with a range from 8.6 at Freising to 23.3 dt/ha at Würzburg, and was higher than the average yield under the STARII with a range from 10.6 at Eichstätt to 27 dt/ha at Würzburg. The average yield for

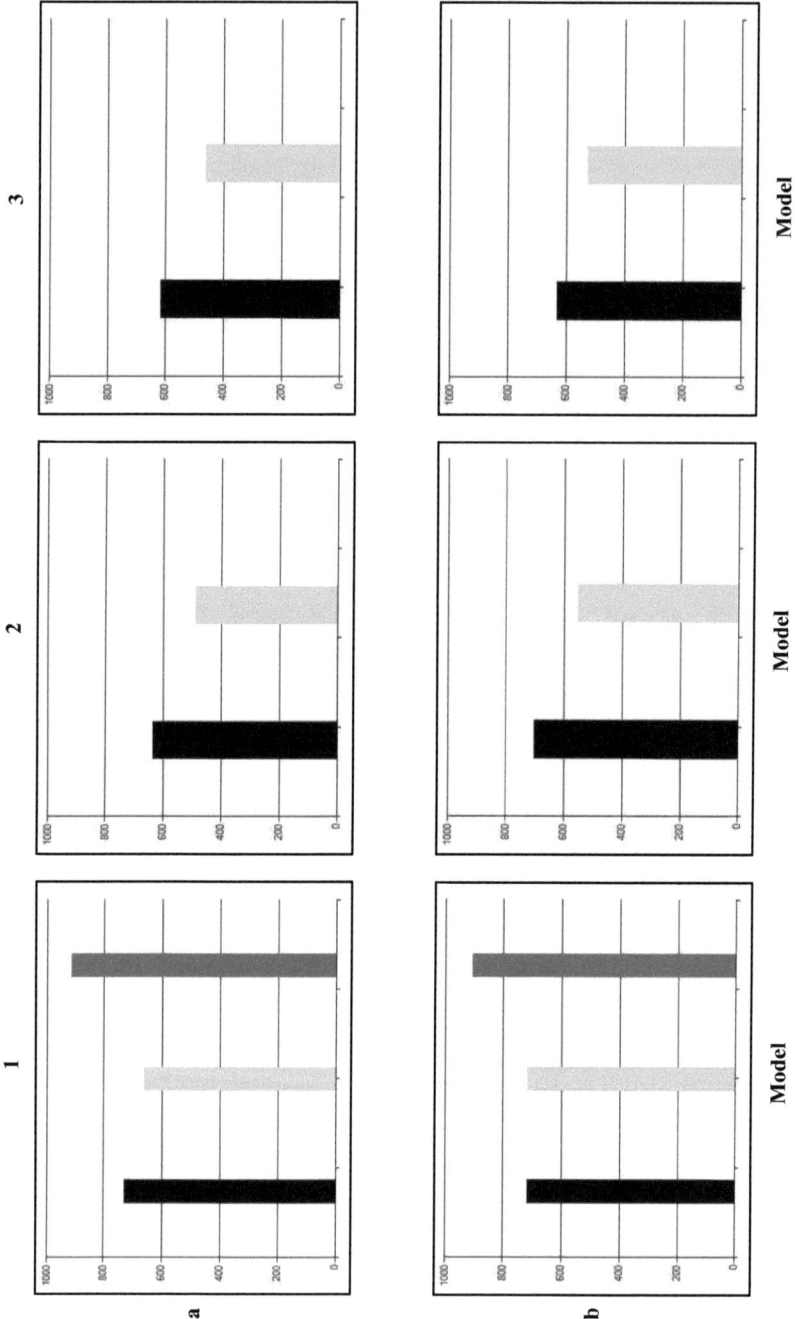

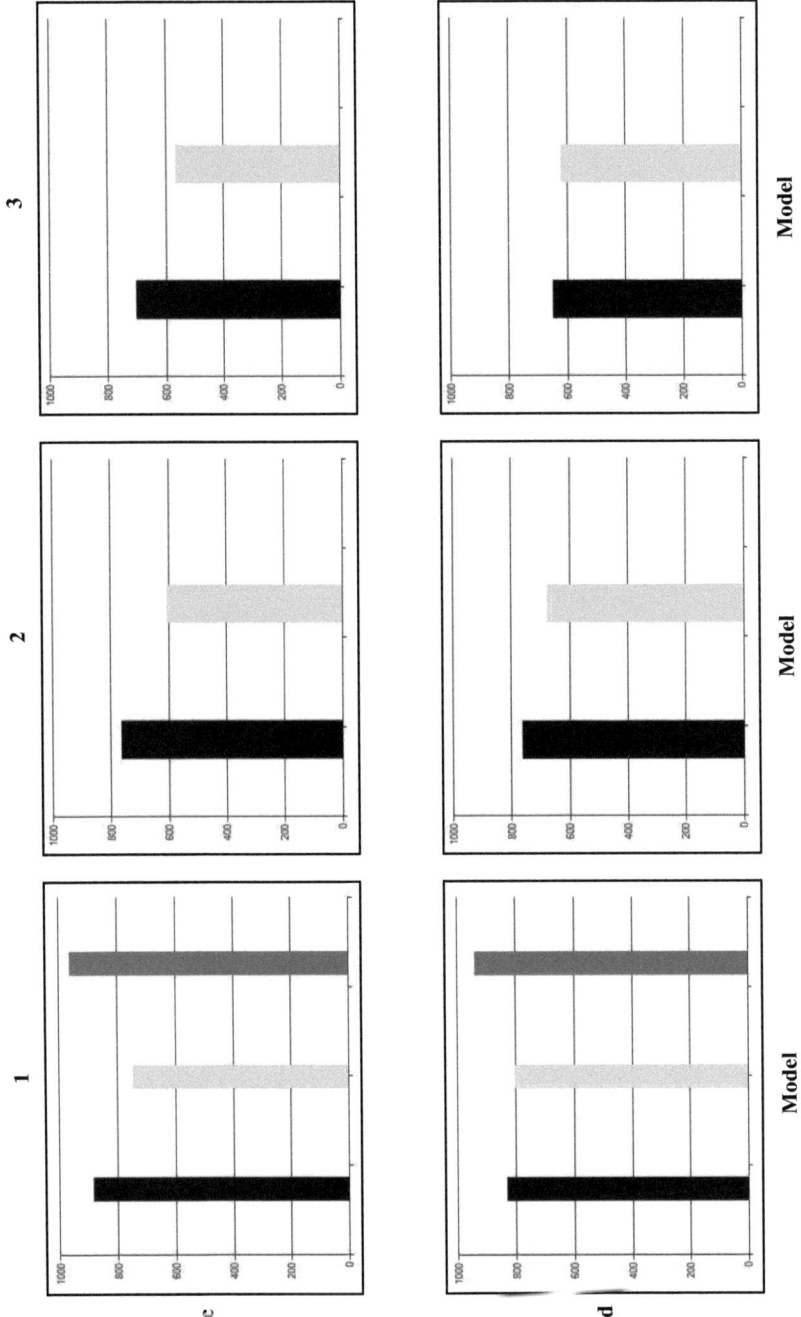

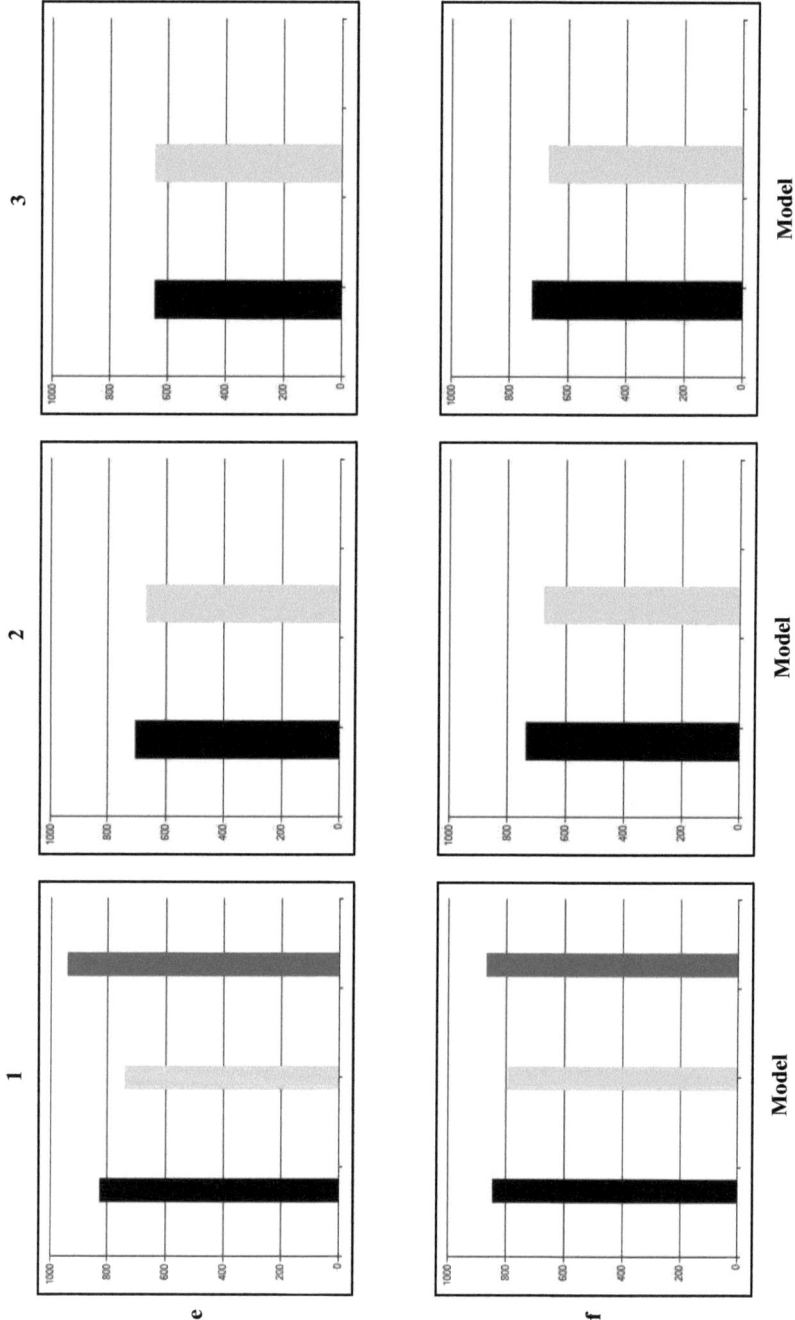

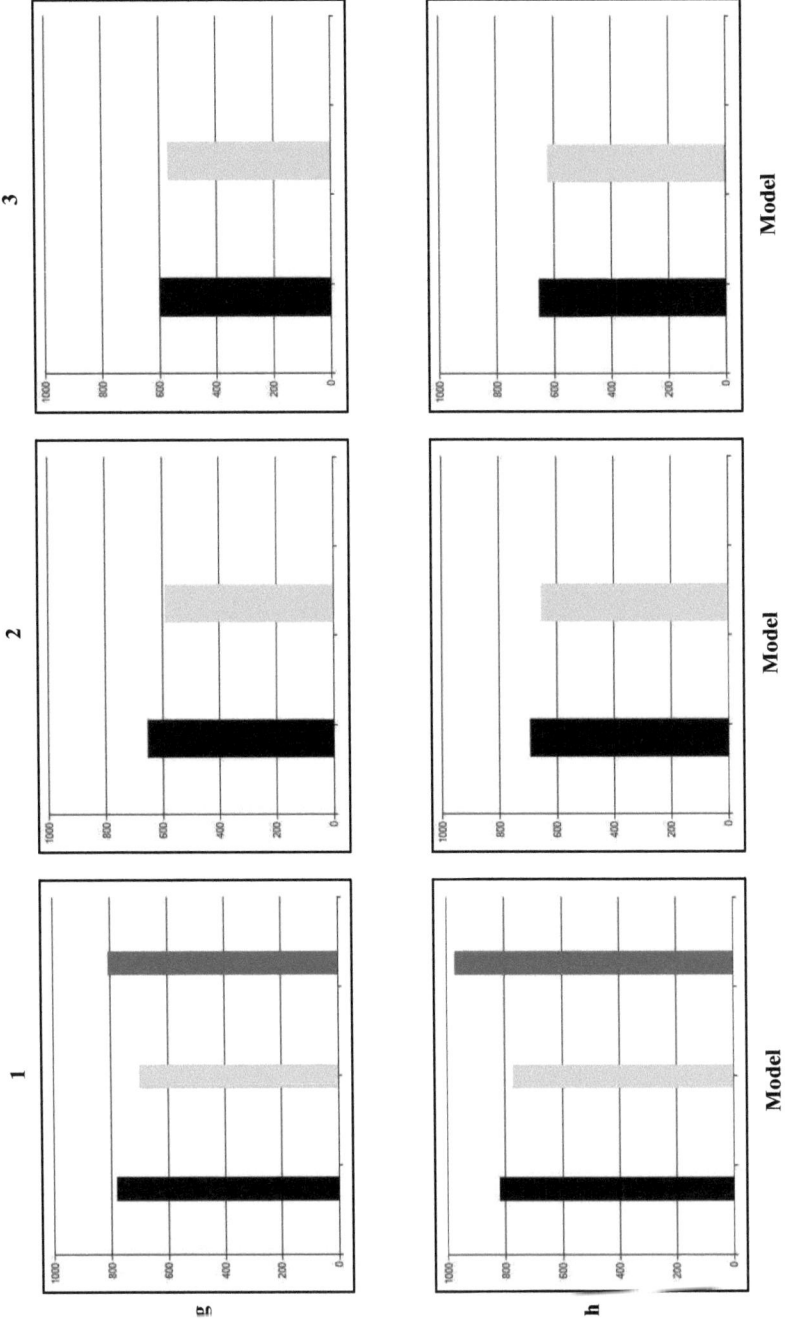

54

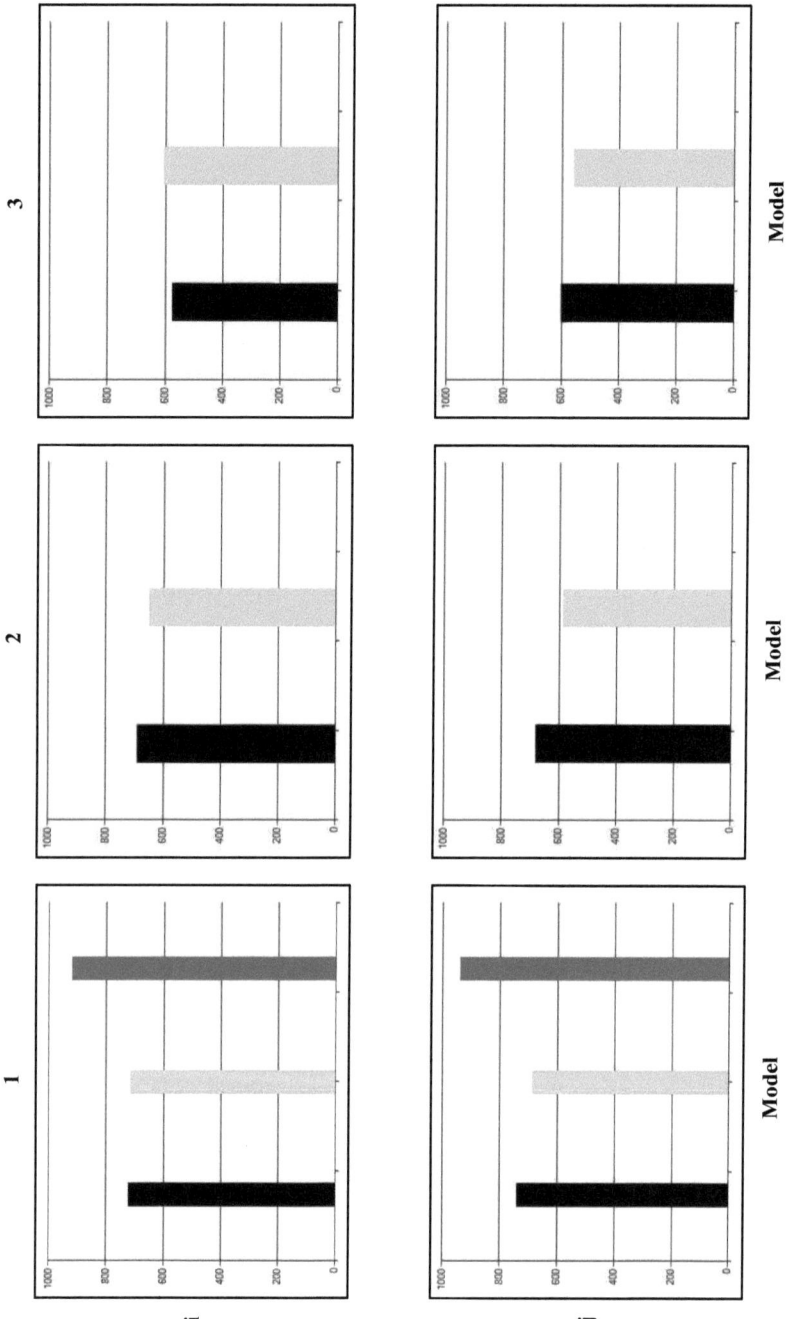

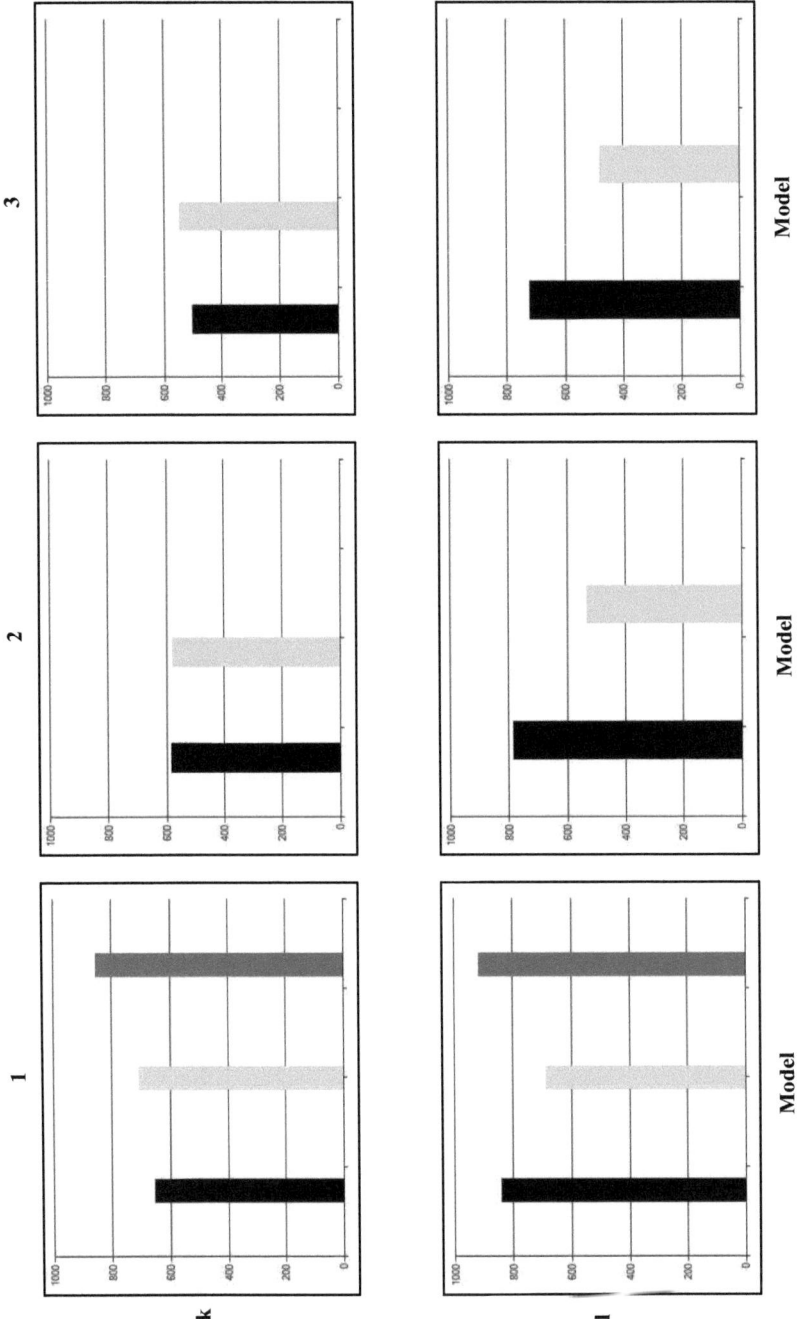

Figure 5. The expected average yield (Dry matter (g.dm/m^2.d)) at the future periods (1) 2021-2050, (2) 2051-2080, and (3) 2071-2100 for the models REMO (■), CLM ( ), and STARII( ) at Würzburg (a), Donau-Ries (b), Freising (c), Passau (d), Regensburg (e), Lichtenfels (f), Eichstätt (g), Landshut (h), Neumarkt (i), Weissburg-Gunzenhausen (j), Main-Spessart (k), and Günzburg (l).

Figure 6. The expected average grain yield at the future periods (a) 2021-2050, (b) 2051-2080, and (c) 2071-2100 for the models REMO (▋), CLM ( ), and STARII( ) at:
1. **Donau-Ries**

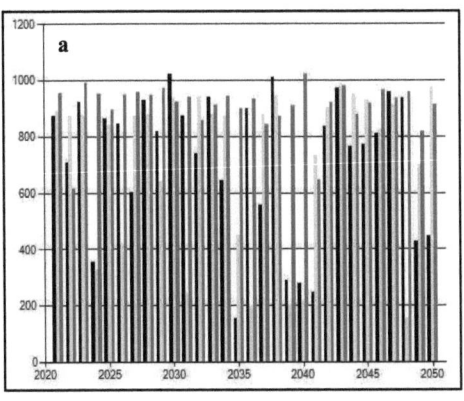

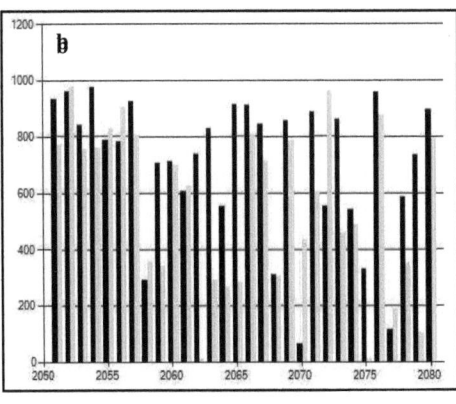

Dry matter (g.dm/m². d)

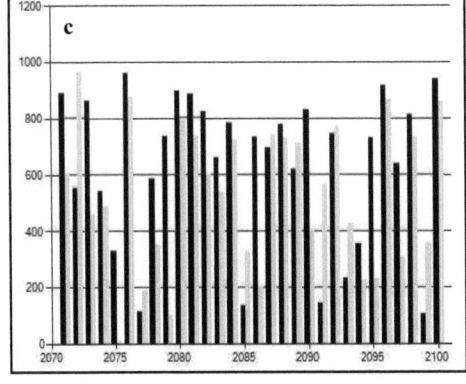

Years

2. Eichstätt

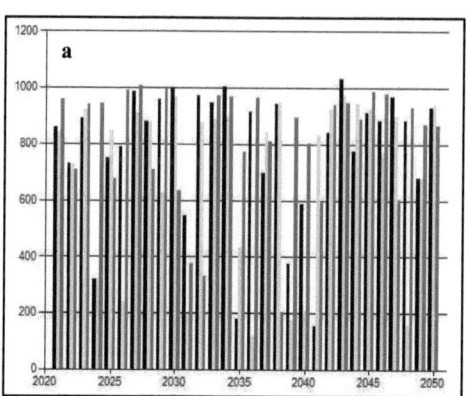

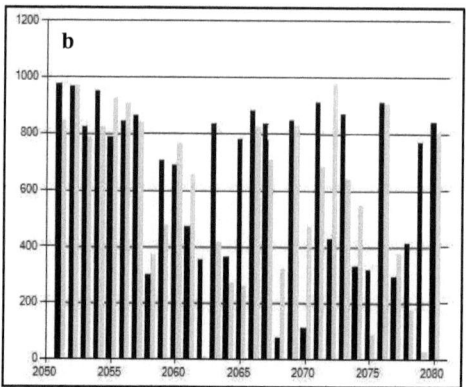

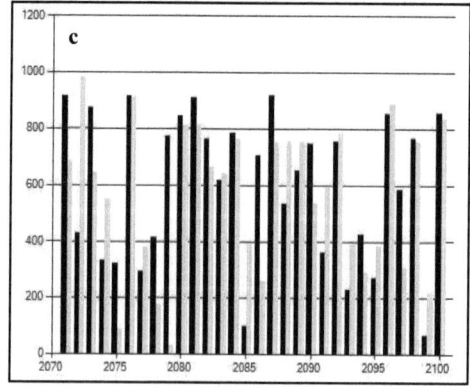

Dry matter (g.dm/m².d)

Years

3. Freising

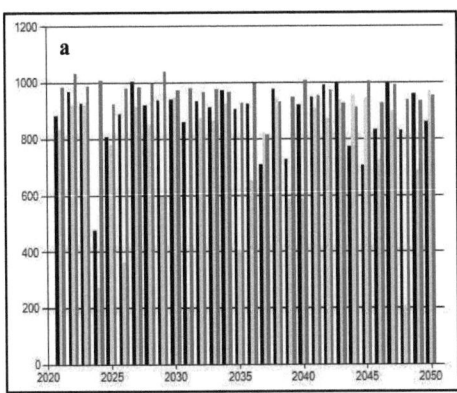

Dry matter (g.dm/m².d)

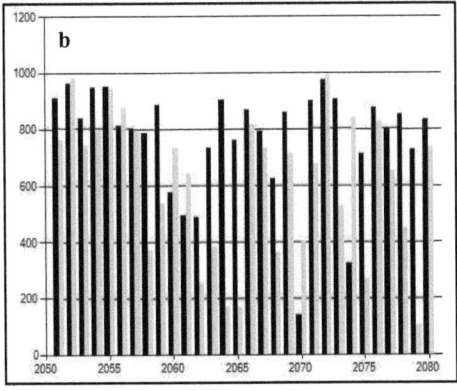

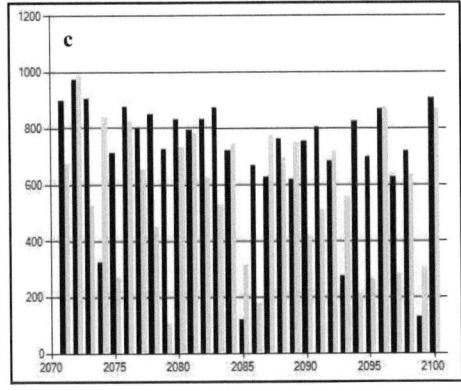

Years

4. Günzburg

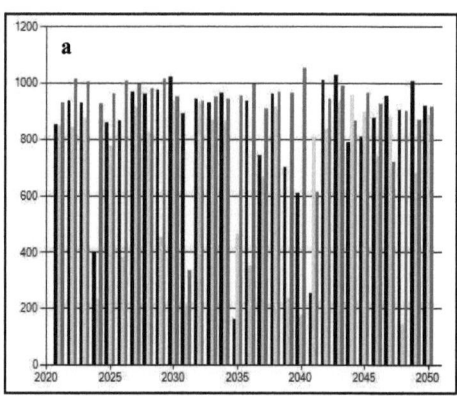

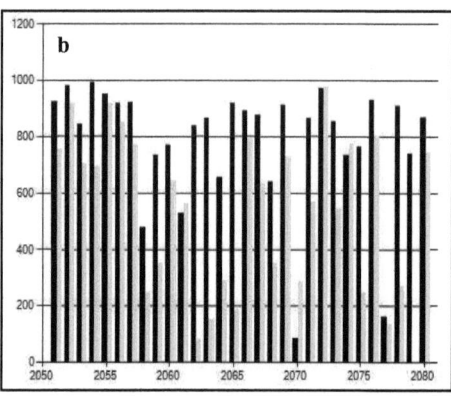

Dry matter (g.dm/m².d)

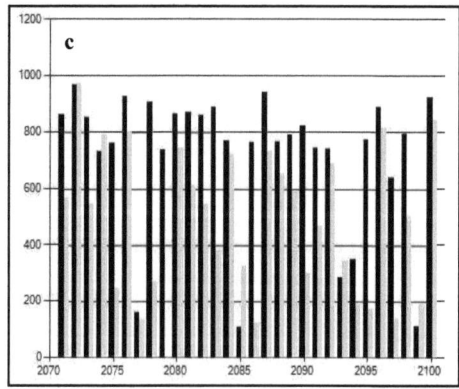

Years

5. Landshut

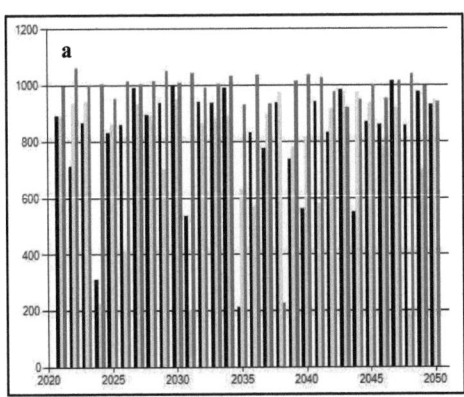

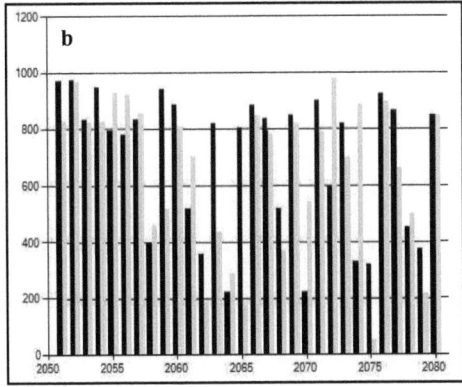

Dry matter ($g.dm/m^2.d$)

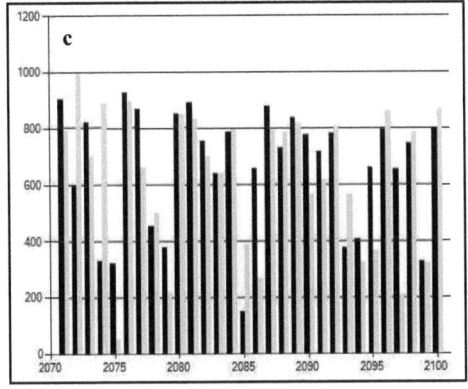

Years

6. Lichtenfels

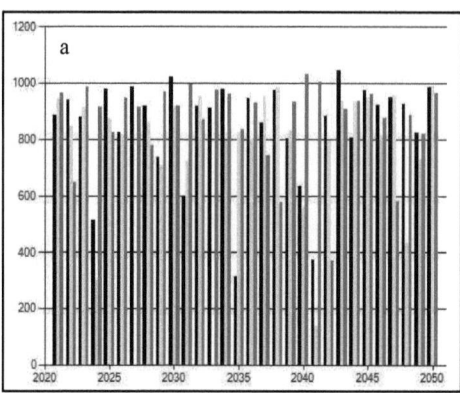

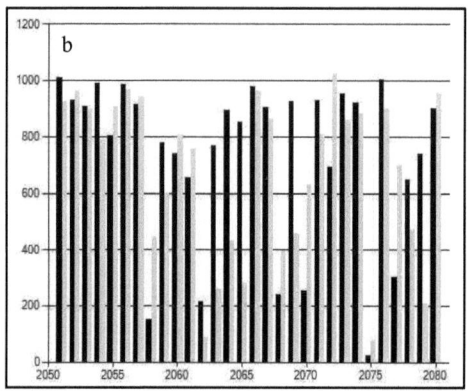

Dry matter (g.dm/m².d)

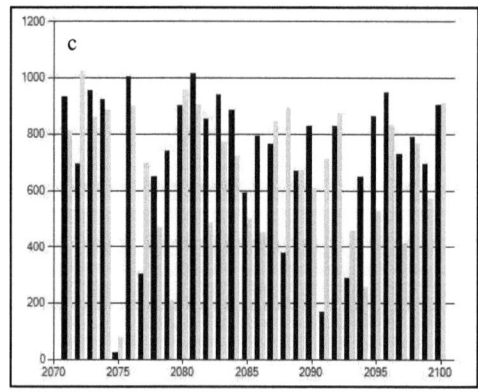

Years

7. Main-Spessart

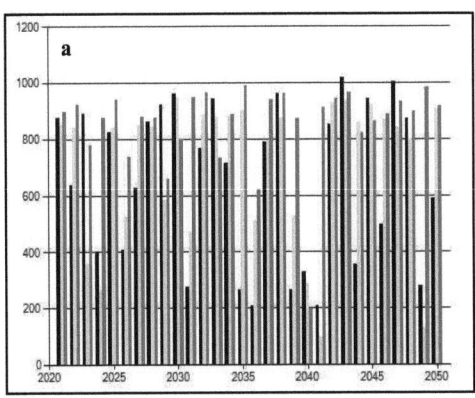

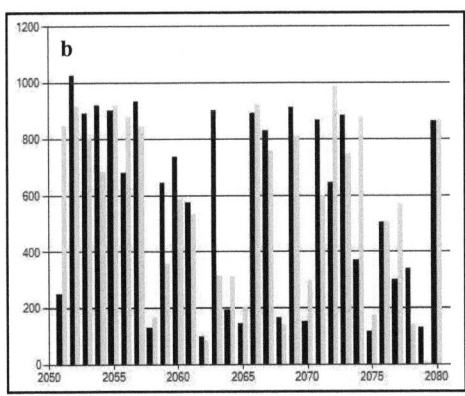

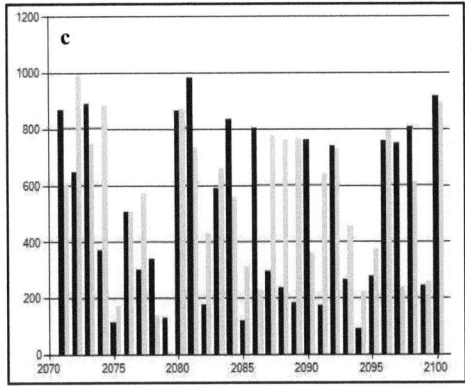

Dry matter (g.dm/m².d)

Years

8. Neumarkt

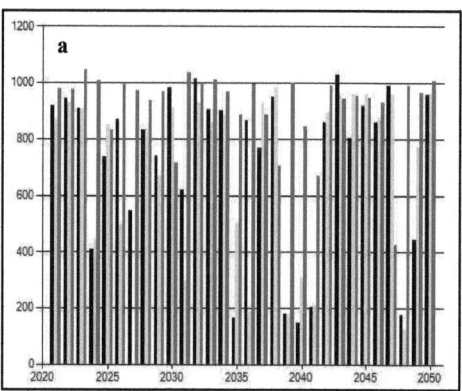

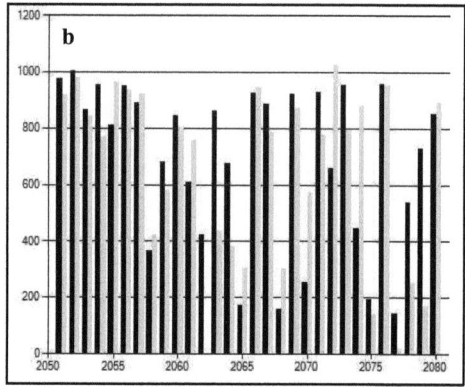

Dry matter (g·dm/m²·d)

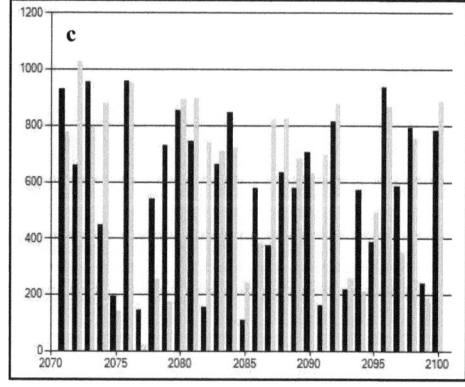

Years

9. Passau

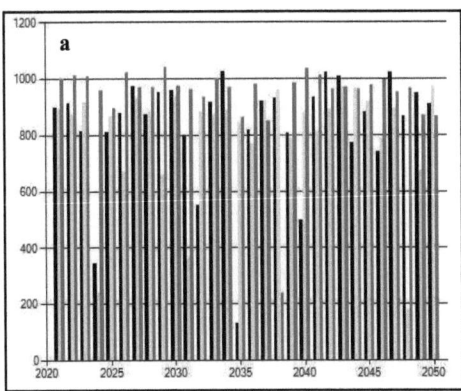

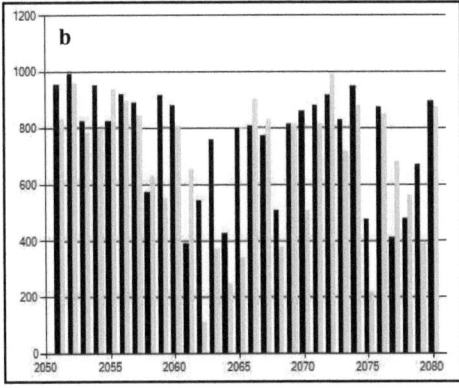

Dry matter (g.dm/m².d)

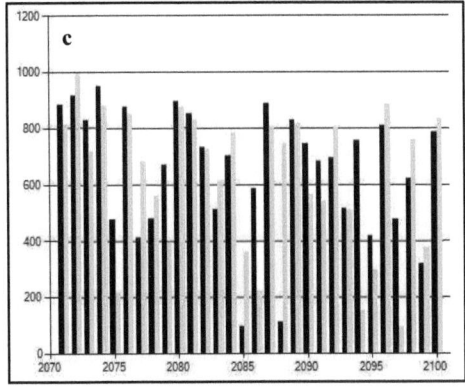

Years

10. Regensburg

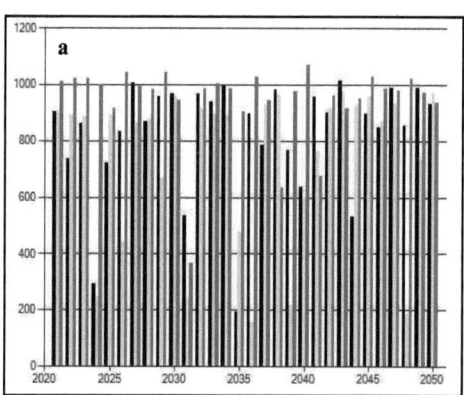

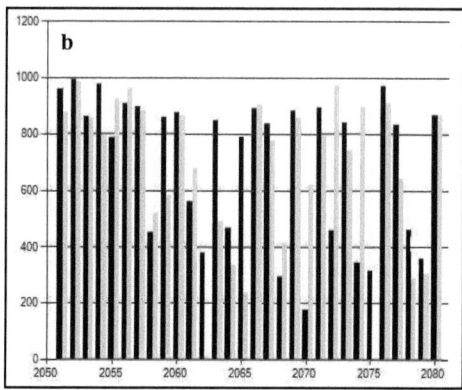

Dry matter (g.dm/m^2.d)

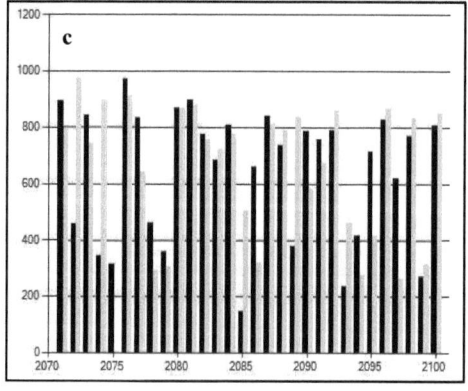

Years

11. Weißenburg-Gunzenhausen

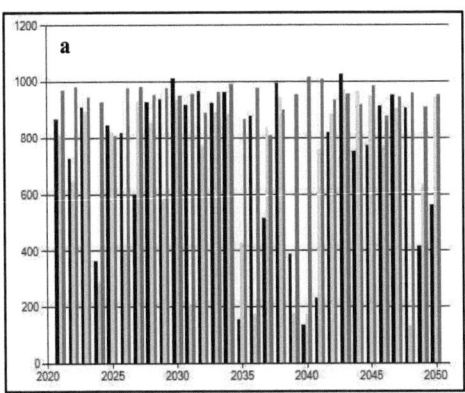

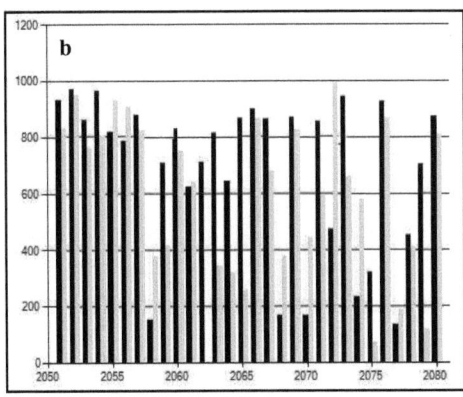

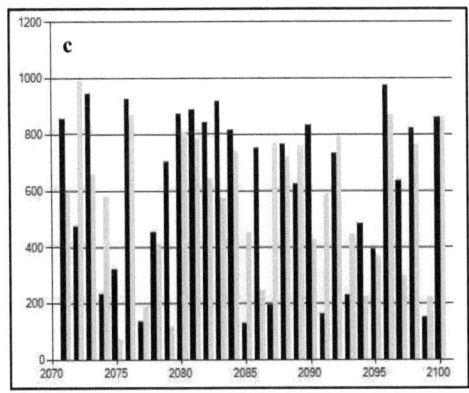

Dry matter (g.dm/m².d)

Years

12. Würzburg

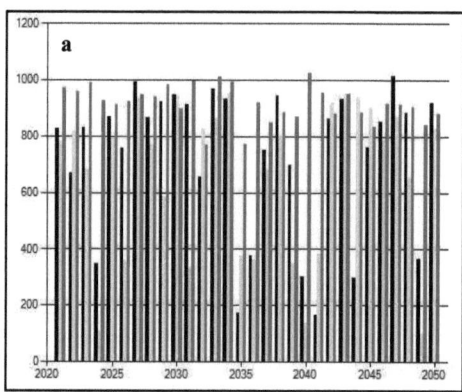

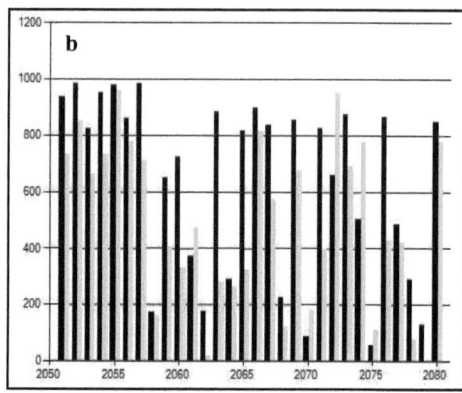

Dry matter (g.dm/m².d)

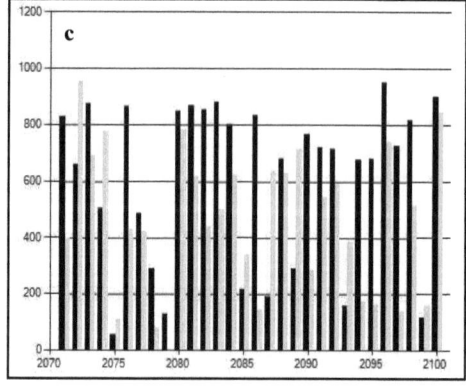

Years

the first period by using STARII was from 90 to 100 dt/ha, except for Landshut (h), which exceeded slightly the 100 dt/ha, and for Eichstätt (g), which was slightly higher than 80 dt/ha. The other two models REMO and CLM data had covered the three studied periods. In many cases, the crop model gave higher average yield for the REMO data than for CLM, which revealed more stress cases during the future seasons. Normally, the simulated higher average yield by REMO was not significant compared to the CLM data. Significant differences by using the two models occurred at Würzburg (a) at the second and third periods with values of 12.5 and 17.9 dt/ha, respectively, in Freising (c) at the three future periods with differences of 10.8, 15.8, and 11 dt/ha, respectively, Passau (d) at the second period with 9.5 dt/ha difference, Günzburg (l) at the three future periods with values of 14.3, 28, and 23.9 dt/ha, respectively, and Weissenburg-Gunzenhausen (j) at the second period with a value of 12.1 dt/ha. The average yield under the CLM data showed a higher value than REMO at Main-Spessart (k) at the first and the third periods with differences of 5.6 and 4.6 dt/ha, respectively, and at Neumarkt (i) at the third period with a value of 3.9 dt/ha.

The individual yields for each of the studied future period fluctuated from high to low during the different seasons per each future period as shown in Figure 6. At the first period (2021 – 2050), the yield exceeded the 100 dt/ha in very few seasons, which exceeded this value with a small amount except for Landshut (h) and Regensburg (e), where almost half of the seasons at the period (2021 – 2050) showed an increase over the 100 dt/ha yield and reached sometimes to 110 dt/ha. The higher yields especially over 100 dt/ha were all from using the future weather data of the model STARII and sometimes REMO, but by using the CLM future weather data, the yields did not reach to 100 dt/ha. Most of the yields at the future period (2021 – 2050) fluctuated from 80 – 100 dt/ha at many of the studied locations, while some locations like Main-Spessart (k), Weissenburg-Gunzenhausen (j), and Würzburg (a) had lower ranges, where most of them were distributed from 60 – 100 dt/ha. In general, with the STARII model, the crop model showed at almost all the locations for the first studied future period (2021 – 2050) insignificant differences among the seasons for each location except for some locations like Eichstätt (g), Günzburg (l), and Main-Spessart (k), which showed some significant differences among the seasons. The crop model revealed at some sites the highest yield under STARII compared to the first period of the other two models REMO and CLM like at Freising (a), Landshut (h), and Neumarkt (i), and at the other sites it displayed higher yields at many seasons of the period. The hot episodes that faced the model at the first period at STARII model had values from 16 to 371 at the whole period,

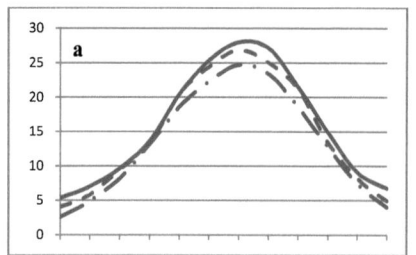
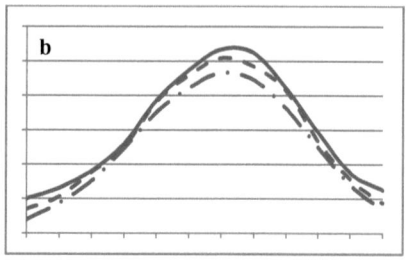
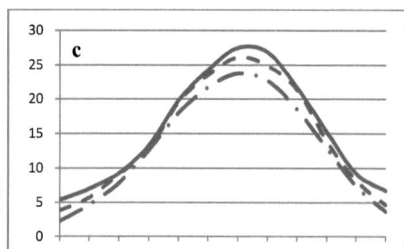
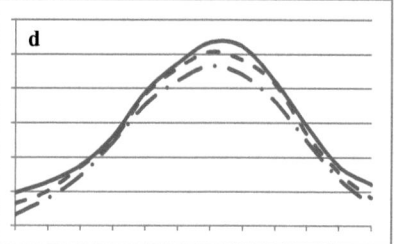
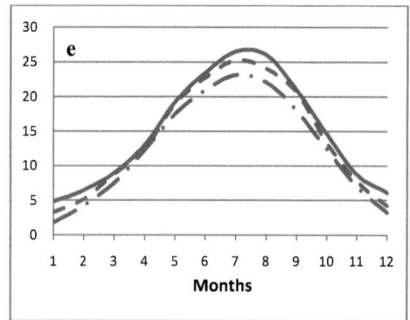
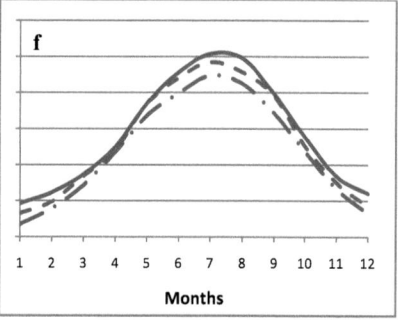

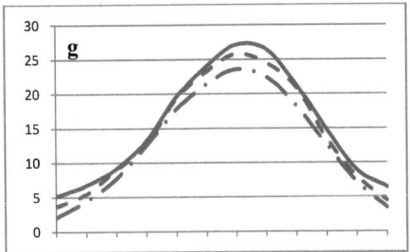
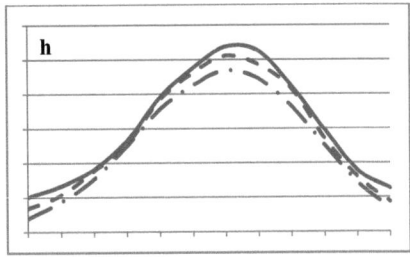
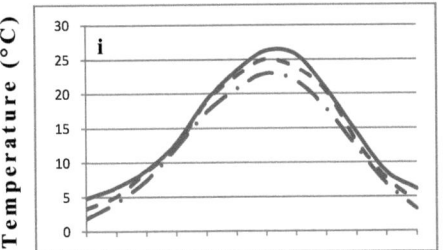
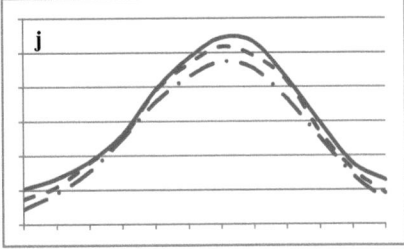
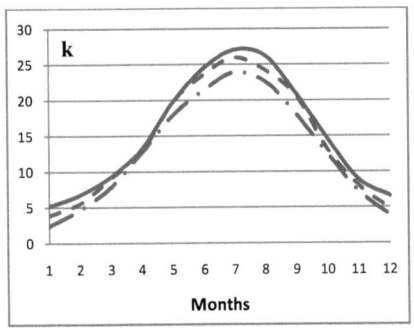
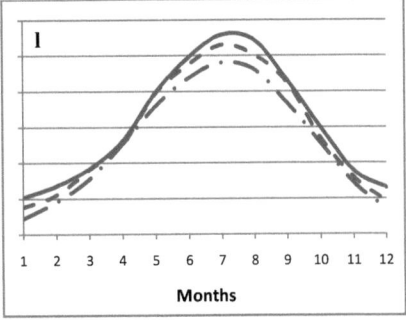

Figure 7. Comparison between the monthly average maximum temperature between the future predicted periods 2021 – 2050 (—·—), 2051 – 2080 (– – ·), and 2071 – 2100 (——) at Würzburg (a), Donau-Ries (b), Freising (c), Passau (d), Regensburg (e), Lichtenfels (f), Eichstätt (g), Landshut (h), Neumarkt (i), Weissburg-Gunzenhausen (j), Main-Spessart (k), and Günzburg (l) for CLM model.

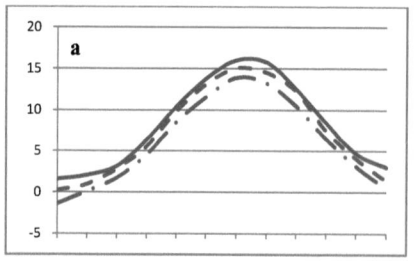
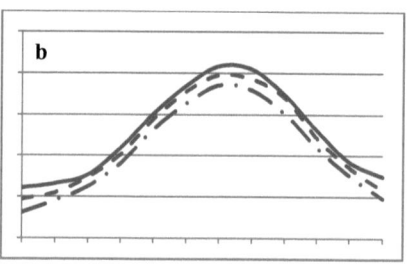
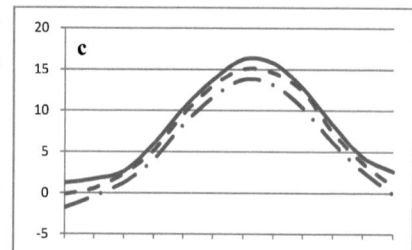
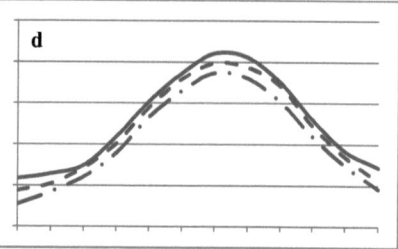
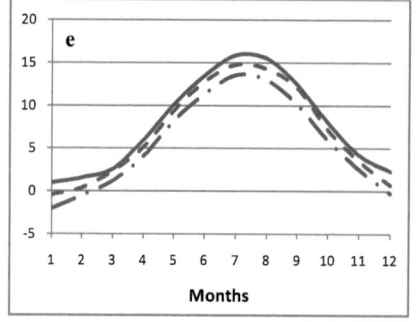
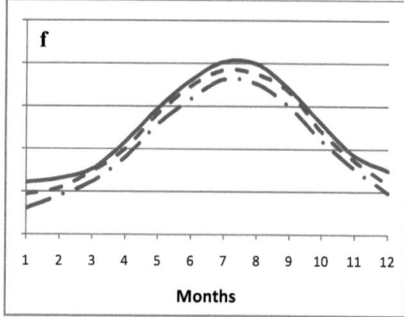

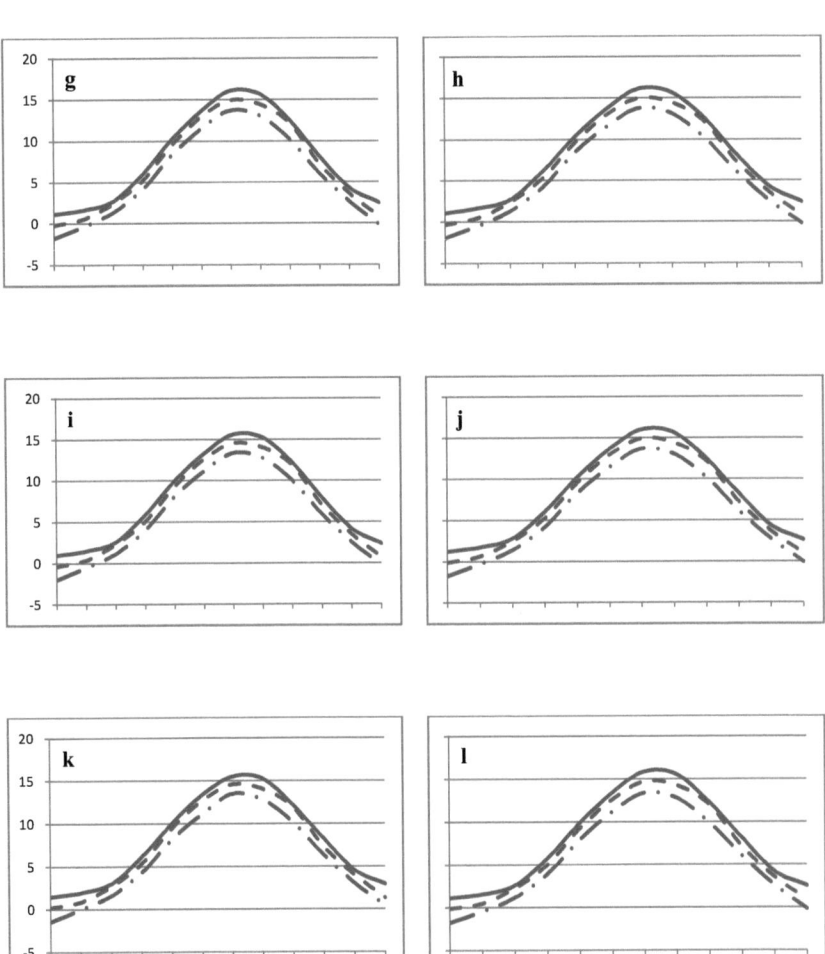

Figure 8. Comparison between the monthly average minimum temperature between the future predicted periods 2021 – 2050 (—·—), 2051 – 2080 (– – ·), and 2071 – 2100 (——) at Würzburg (a), Donau-Ries (b), Freising (c), Passau (d), Regensburg (e), Lichtenfels (f), Eichstätt (g), Landshut (h), Neumarkt (i), Weissburg-Gunzenhausen (j), Main-Spessart (k), and Günzburg (l) for CLM model.

where the lowest value were in Günzburg, which faced 16 hot days with an average of 0.5 hot days/year, and the highest value 371 appeared in Main-Spessart with an average of 12.4 hot days/year. The STARII crop model had predicted at some sites lower hot episodes at the first period as in Günzburg, Landshut, Neumarkt, and Regensburg, which faced 16, 42, 55, and 69 hot days, respectively, with an average of 0.5, 1.4, 1.8, and 2.3 hot days/year, respectively, while the sites, which had higher hot episodes, were Main-Spessart, Würzburg, Eichstätt, and Lichtenfels with hot days of 371, 285, 202, and 198, respectively, with an average of 12.4, 9.5, 6.7, and 6.6 hot days/year, respectively. The hot episodes, preheated by the models REMO and CLM during the first period, were higher than for STARII. The model confronted with REMO during the first period hot episodes in a range from 117 hot days with an average 3.9 hot days/year in Neumarkt to 274 hot days with an average of 9.1 hot days/year in Passau. As well as the model with CLM future weather data encountered during the first period hot episodes in a range from 185 hot days with an average of 6.2 hot days/year in Lichtenfels to 501 hot days with an average of 16.7 hot days/year in Würzburg.

Extremely low yield seasons, which ranged here from 40 to 10 dt/ha, appeared by using the three future periods. With REMO and CLM revealed the crop model during the future periods few to many extremely low yield seasons for all the studied sites, where as with STARII data very few extremely low yield seasons occurred in Eichstätt (g), Landshut (h), Passau (d), and Regensburg (e), which indicated 2, 1, 1, and 1 extremely low yield seasons, respectively. The appearance of the extremely low yield was repeated by using the REMO and CLM data at all the studied sites and was higher than for the STARII first period. In general, the crop model showed by using CLM data more extremely low yield seasons during each studied future period compared to REMO data, where the occurrence of these seasons were sometimes low at the first period (2021 – 2050) by both models at some sites like Landshut (h), Lichtenfels (f), and Passau (d) with values of 2 seasons under REMO future weather data, and with values of 3, 1, and 3 seasons, respectively, under CLM data. These extremely low yield seasons continued having low appearance only under the REMO second future period data for the sites Günzburg (l), Main-Spessart (k), Freising (c), and Regensburg (e), with 3, 3, 2, and 2 seasons, respectively. Otherwise the first future period showed also a relatively high extremely low yield season occurrence as in Neumarkt (i), Main-Spessart (k), and Würzburg (a), which indicated eight extremely low yield seasons for each by using the CLM data, and in Würzburg (a) with the REMO data, the crop model indicated nine extremely low yield seasons. At the second and third future periods the crop model showed higher extremely low yield seasons for REMO and CLM with a range from 8 to 13 seasons. The crop model revealed also at the second and third periods higher hot episodes than the first period, where by REMO the hot episodes appeared at the second period in the range between 364 hot days in Lichtenfels with an average of 12.1 hot days/year and 692 hot days in Passau with an average of 23.1 hot days/year, while at the third period the model indicated hot episodes that fluctuated between 420 hot days in Lichtenfels with an average of 14 hot days/year and 846 hot days in Passau with an average of 28.2 hot days/year. The hot episodes under CLM revealed higher values at the second and third periods compared to REMO model, where at the second period, by the CLM the model faced hot episodes that fluctuated from 400 hot days in Lichtenfels with an average of 13.3 hot days/year to 817 hot days in Würzburg with an average of 27.2 hot days/year. The hot episodes by the third period of CLM ranged from 514 hot days with an average of 17.13 hot days/year to 966 hot days in Würzburg with an average of 32.2 hot days/year.

Thus, the second future period (2051 – 2080) was characterized for all the studied locations with higher yield from 100 to 80 or at some sites from 100 to 60 dt/ha at the earlier seasons (8 – 10), then the yields' distribution of the other seasons fluctuated from very high yields almost at 100 dt/ha till very low at about 10 dt/ha. And so on, continued at the third future period (2071 – 2100), the yield fluctuated from high values, which reached to 100 dt/ha, to low values, which decreased to 10 dt/ha or lower, but with more frequency of the low yield seasons. Normally, the frequency of the extremely low yield seasons at the third future period was more than at the second one, except for at Lichtenfels (f) under the CLM data, where the model showed two extremely low yield seasons at the third future period but four at the second one. The model by using the CLM data showed always a higher appearance of the low extreme events from extremely low yield and no yield seasons (0 – 5 dt/ha) than by REMO, although the model yields under CLM data were not all the time lower than under REMO data. The occurrence of the no yield (0 – 5 dt/ha) case were revealed at one or two seasons at the second and third future periods at some sites of the studied ones. The most frequent seasons, which showed the no yield cases, were 2061/62, 2074/75, and 2078/79. The first season (2061/62) indicated the no yield at the sites Donau-Ries (d), Eichstätt (g), and Würzburg (a), which were confronted with 45, 45, and 67 hot days, respectively, by the CLM data. As well as the season 2078/79 showed only by the CLM data no yield at some sites, which were Freising (c), Günzburg (l), Main-Spessart (k), Neumarkt (i), and Würzburg (a), with values of 30, 37, 33, 24, and 36 hot days at that season, respectively. The season 2074/75 indicated also no yield prediction once by the REMO data at Main-Spessart (k) with 32 hot days, and once by the CLM at Regensburg (e) with 40 hot days.

The predicted temperature at the second and third future periods increased more than for the first one for the REMO and CLM (Figure 7 and 8, Appendix A2 and A3). For REMO, the maximum temperature increased during the different months and locations of the second future period compared to the first one with ranges from 0.35°C in April at Main-Spessart to 3 °C in September at Freising, and the minimum temperature increased from 0.79°C in February at Passau to 2.4°C in September at Günzburg. The maximum and minimum temperatures increased at the third future period more than in the first one with ranges from 1.19°C in April at Main-Spessart to 4.1°C in August at Passau, and from 1.26°C in March at Landshut to 3.6°C in January at Main-Spessart, respectively. With the same way for CLM, the increases of the maximum and minimum temperatures, of the second future period compared to the first one, were at the ranges from 0.04°C in April to 2.8°C in September both at Günzburg, and from 0.86°C in February at Main-Spessart to 2.09°C in September at Günzburg, respectively, and the increases of the maximum and minimum temperatures of the third future period compared to the first one were from 0.45°C in April at Günzburg to 4.19°C in August at Freising, and from 1.37°C in March at Landshut to 3.1°C in January at Passau, respectively. Therefore, the occurrence of the cases of the extremely low or no yields at the third future period was higher than the second one, depending on the higher temperature increase at the third period compared to the second one. Also, the highest maximum temperature during the different studied sites and seasons was higher at the third period than the second one for the REMO and CLM, where at REMO, the highest maximum temperature at the second period for all the studied sites was at the range from 40.3° at 2/9/2062 in Donau-Ries to 43.2° at 18/6/2061 in Freising, and it varied in the third period within the range from 39°C at 8/8/2086 in Lichtenfels to 46.5°C at 30/7/2091 in Passau. At CLM, the highest maximum temperature for all the studied sites at the second and third future periods was at the ranges from 40.4°C at 28/6/2070 in Freising to 44.8°C at 17/7/2078 in Würzburg, and from 42.2°C at 5/7/2099 in Passau to 46.5°C at 4/7/2094 in Würzburg, respectively.

3.5 Spatial yield distribution in Bavaria

The Maps 2 – 7 are representing the spatial yield distribution, appearance of the extreme temperature during the growing season and the grain-filling duration of winter wheat in Bavaria for the weather models REMO and CLM for the three future periods (2021 – 2050, 2051 – 2080, and 2071 – 2100).

The simulated yields for Bavaria in the first future period for CLM on Map 5 (a) ranged from 57 to 90 dt/ha, where the lower yield appeared mainly in the northwest (Unterfranken except for its most western part) and the lower middle (the northern half of Schwaben and Oberbayern) of Bavaria, and the higher yield appeared in the east and northeast (east of Oberfranken, Oberpfalz and Niderbayern) and south (south of Schwaben and Oberbayern). In the second future period, the simulated yield ranged from 41 to 89.6 dt/ha (Map 5 (b)), where the distributed lower yield area increased to cover almost all Unterfranken (northeast of Bavaria) and the lower middle (the north half of Schwaben and Oberbayern and the south half of Niederbayern) of Bavaria, but in the third future period (Map 5 (c)), the estimated high yield was only found in the northeast, the easternmost and the southernmost of Bavaria with ranges of 80 – 90 dt/ha, while the rest parts had low yields ranging from 30 to 75 dt/ha. With the same concept was estimated the average grain-filling duration period during the three future periods for the CLM model (Map 7 (a, b and c)), where the longest grain-filling duration appeared in the southernmost part of Bavaria with ranges from 250 to 385; from 200 to 380; and from 150 to 300 days for the three future periods, respectively. Also the northeast and the eastern corner of Bavaria showed a higher grain-filling duration, which ranged from 200 to 250; from 115 to 185; and from 95 to 115 days during the three future periods, respectively, while the other parts of Bavaria (Unterfranken, Mittelfranken, the west of Oberpfalz and Niederbayern, and the northern half of Schwaben and Oberbayern) had a relatively low grain-filling duration ranging from 80 to 200; from 75 to 115; and from 80 to 95 days during the three future periods, respectively. The model estimated a higher appearance of the extreme maximum temperature for the CLM model in the northwest (Unterfranken) and the lower middle (the northern half of Schwaben and Oberbayern and the southern half of Niederbayern) of Bavaria for the three studied future periods (Maps 6 (a, b and c)) with ranges of 10 – 17.5, 14 – 22, and 13 – 19 days, respectively, while the lower appearance of the hot temperature episodes were in the east and northeast (east of Oberfranken, Oberpfalz and Niederbayern) and south (south of Schwaben and Oberbayern) of Bavaria with ranges of 0 – 4, 0 – 6, and 1 – 6 days for the three future periods, respectively. During the first future period of the model simulation extreme maximum temperature occurred more often than higher ones across Bavaria, while the percentage of the higher occurrence of the extreme maximum temperature increased gradually across Bavaria during the second and third future periods till the lowest occurrence (0 – 1 day) was no longer represented within the third future period map of the CLM model. Therefore, the highest occurrence range of the extreme maximum temperature (19 – 22 days) appeared only during the second and third future period, while it did not appear during the first one. Nevertheless, the model revealed the highest occurrence range of the extreme maximum temperature (19 – 22 days) during both, the second and the third future periods of the CLM model. But only the second period showed the whole range, whilst the model of the third one had only a maximum of 19 hot days.

The yield simulated by the REMO model had a different distribution than in the CLM model, where the highest estimated yield and the lower appearance of the hot temperature episodes during the first future period (Map 2 (a)) appeared in Oberfranken, Oberpfalz, the estern corner of Niederbayern, in a part in the north of Unterfranken, Eichstätt in Oberbayern and Donau-Ries in Schwaben with a range of 93.5 – 96 dt/ha and 0 – 2.6 days (Map 3 (a)) respectively, while the northeast and the southern part of Bavaria had a lower yield and higher occurrence of hot days ranging from 87.7 to 93 dt/ha and 3.5 – 5.5 days, respectively. During the second future period, the simulated yield was higher and the number of simulated hot days was low in the southernmost corner, the northeast and the eastern corner of Bavaria with a range of 90 – 94 dt/ha and 0 – 4 days (Maps 2 (b) and 3 (b)), respectively, and with a large range within the yield and number of hot days in the other parts of Bavaria ranging from 73.7 to 93 dt/ha and from 4 to 7.4 days, respectively. For the third future period map on Map 2 (c), the model simulated the winter wheat yield nearly close to the distribution of the first future period but with different values, where the higher yield appeared in the northeast, the eastern corner, the southernmost corner, and a small part in the north of Unterfranken in Bavaria with values ranging from 86 to 91 dt/ha, whereas the southern part of Bavaria had almost the lowest estimated yield with a range of 76.6 – 83 dt/ha. The occurrence of the extreme maximum temperature (Map 3 (c)) was in general low in the north of Bavaria with a range of 2.5 – 3 days and high in the south with a range of 4 – 6.5 days, with small areas in the far south with a range of 0.5 – 2.5 days. The grain-filling duration was estimated to be very long in the southernmost part of Bavaria (Map 4 (a, b and c)) with ranges from 230 to 415; from 160 to 320; and from 125 to 214 days during the three future periods, respectively, whilst the other parts of Bavaria had a lower duration than the last one with ranges of 95 – 180, 92 – 115, and 98 – 135 days during the three future periods respectively.

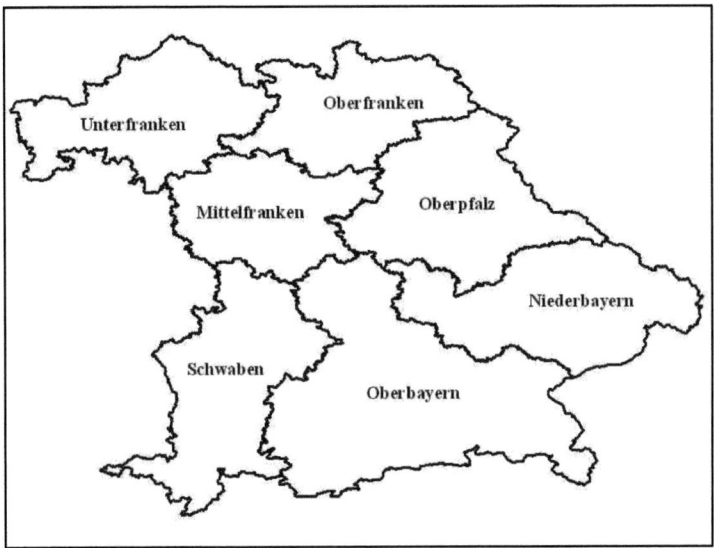

Map 1. The administrative districts of Bavaria

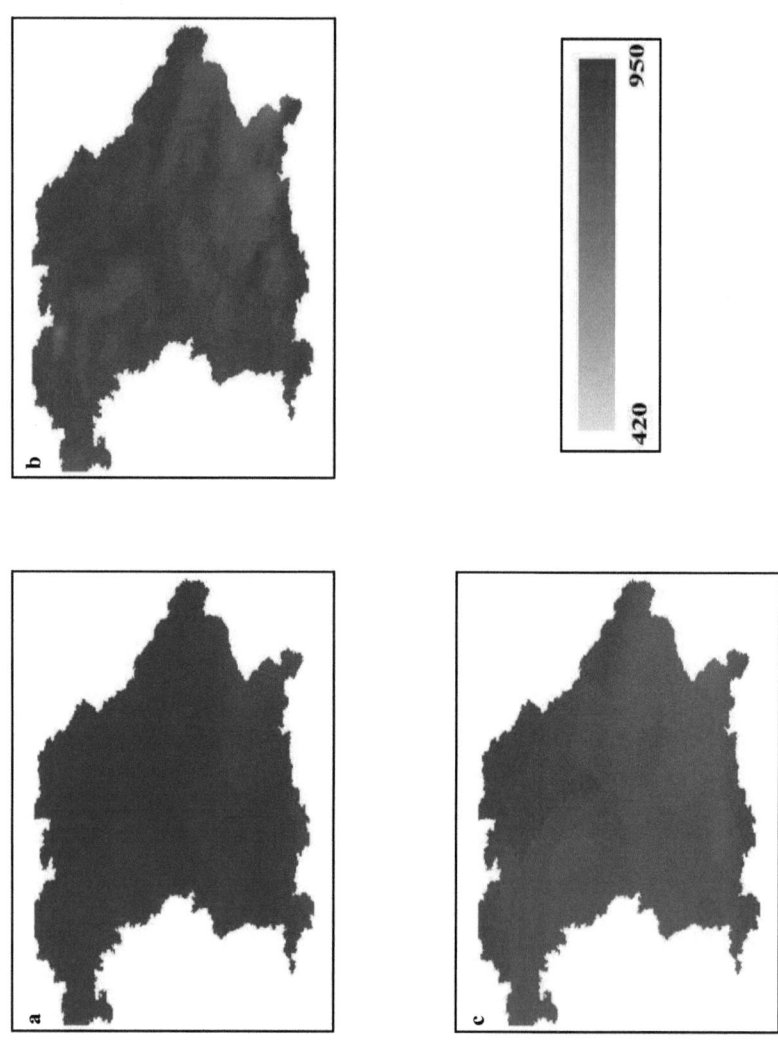

Map 2. The estimated wheat yield distribution in Bavaria by using the REMO model at the first (a), second (b) and third (c) future period.

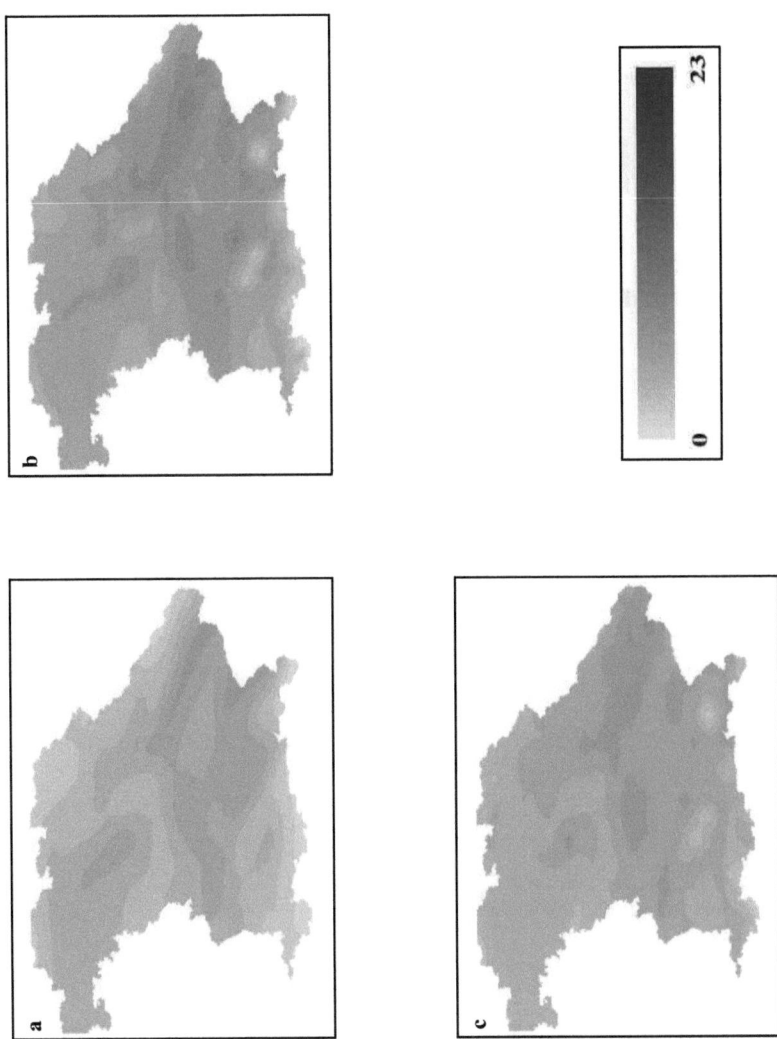

Map 3. The distribution of the estimated number of hot episodes in Bavaria for the REMO model at the first (a), second (b) and third (c) future period.

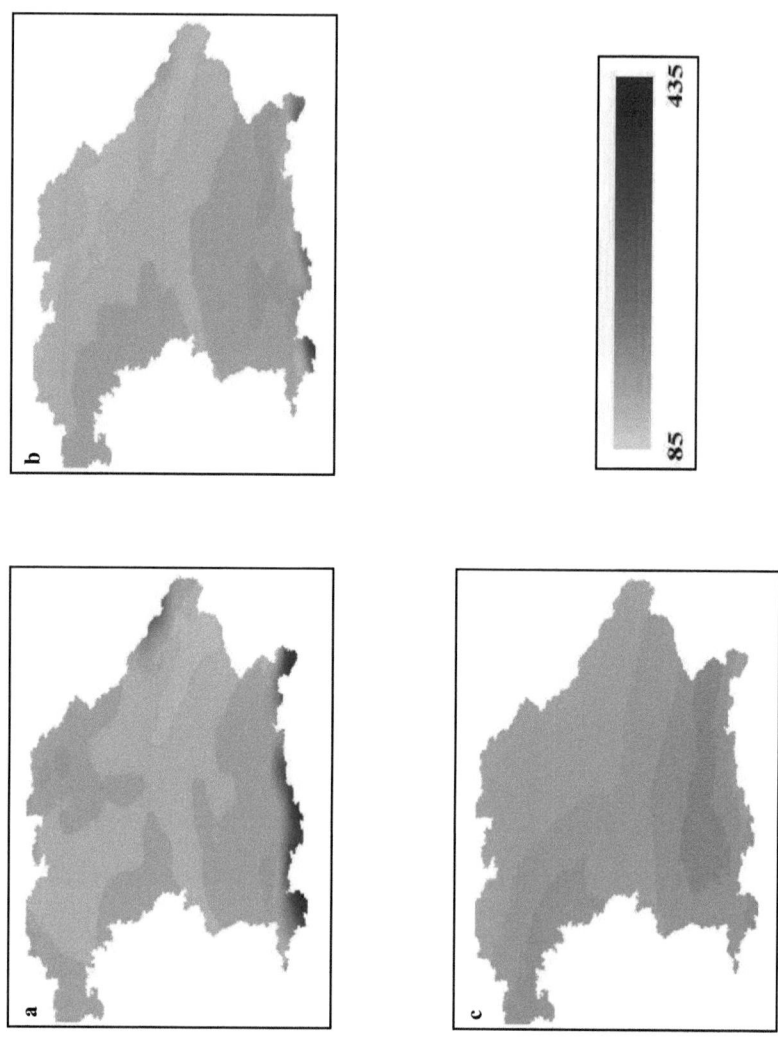

Map 4. The distribution of the estimated grain-filling duration in Bavaria for the REMO model at the first (a), second (b) and third (c) future period.

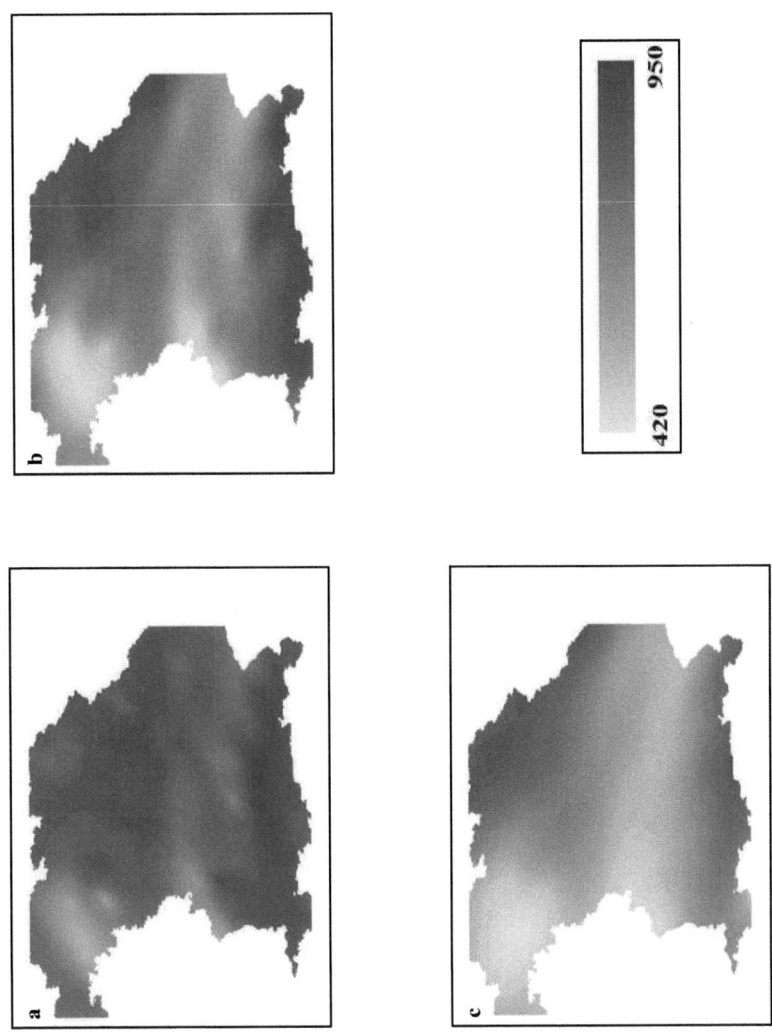

Map 5. The estimated wheat yield distribution in Bavaria by using the CLM model at the first (a), second (b) and third (c) future period.

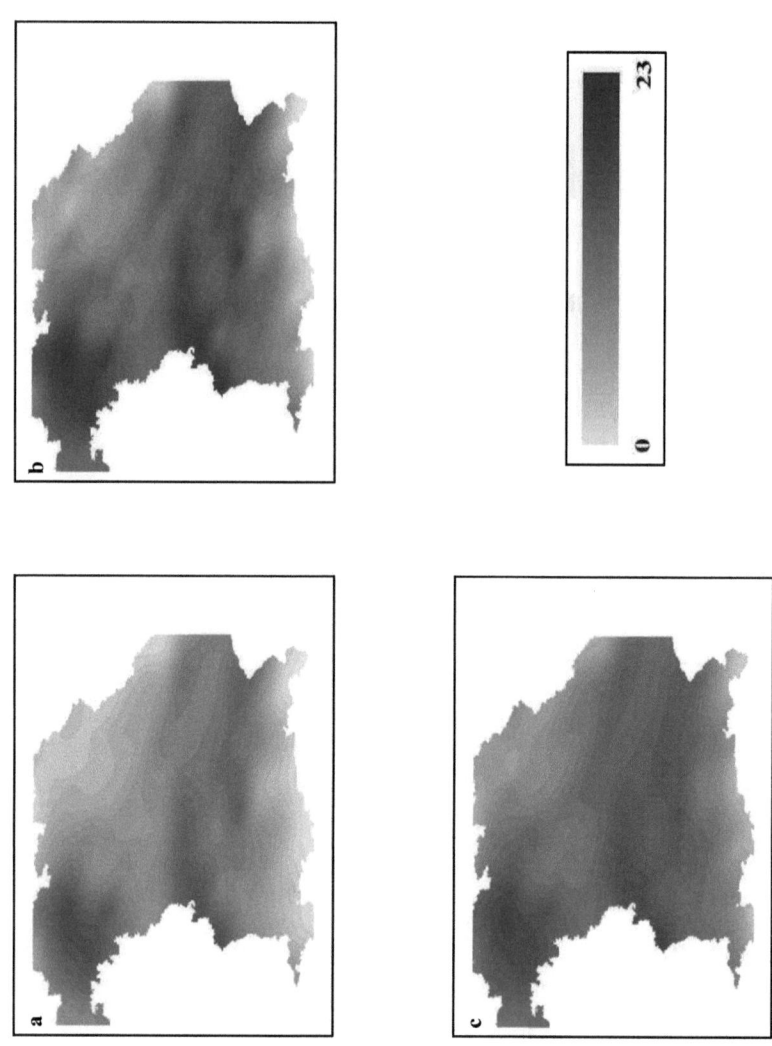

Map 6. The distribution of the estimated number of hot episodes in Bavaria for the CLM model at the first (a), second (b) and third (c) future period.

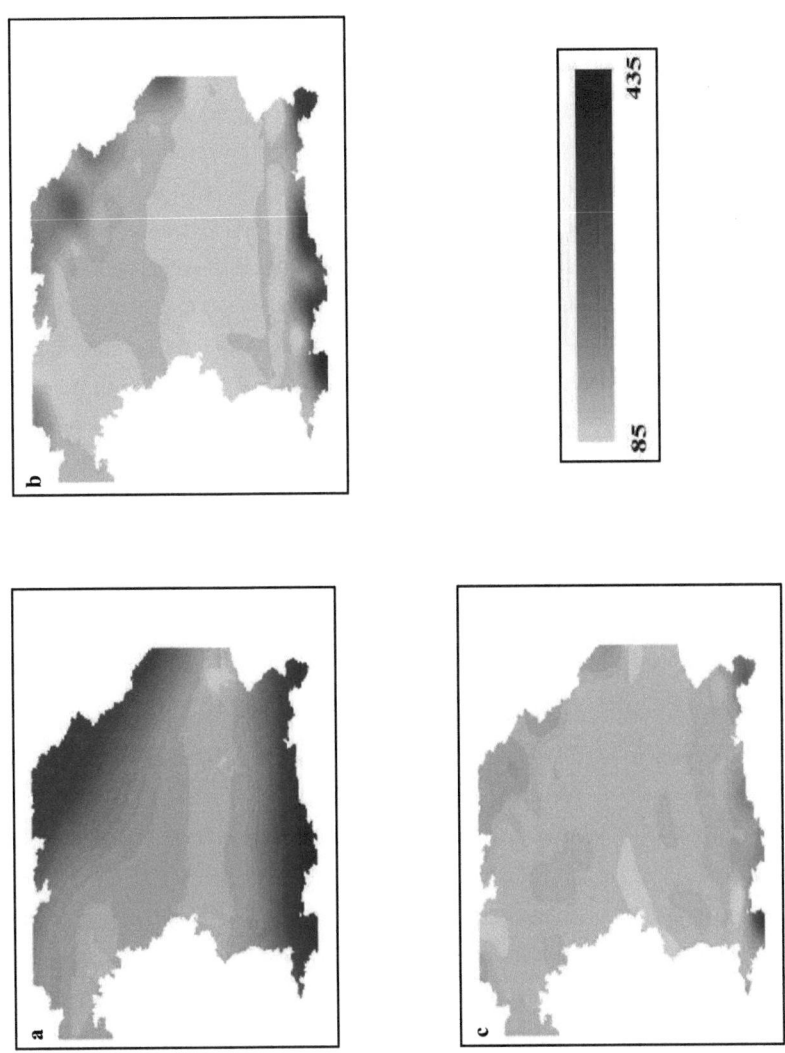

Map 7. The distribution of the estimated grain-filling duration in Bavaria for the CLM model at the first (a), second (b) and third (c) future period.

3.6 The computer-based model description

The computer based model had been developed by using the programming language Microsoft Visual Basic.NET©, where the required plant and weather data for the model had been organized in a database using the database engine Microsoft SQL Server©. The application model started from the Photo 1, where the user will select first the required city, which after selecting it, the available sites in the model database for that selected city will be displayed at the nearby list box, and also the location of that selected city will be marked at the Bavarian map up on the left side. The second option at Photo 1 is selecting with weather data the user wants to apply, by selecting one of the radio buttons for the current or future weather. The default selected option is the current weather, which display all the available current weather data of the selected site from the database, represented in list boxes; year, month, and day, for selecting the planting date. The last part of the model form is the soil attributes data, which requires manual inputs from the user. The soil required data is the electrical conductivity of the soil (EC), the soil bulk density (g/cm3), the organic carbon percent in the soil (%), the sand and clay percent in the soil (%) at 3 different depths (40, 80, and 120 cm). The quality control is applied on the model form, where it detects any empty item or false written item in the form and gives an alert of it without submitting the form.

After selecting and writing the required data from the user at the last form and submitting the form, another page will be opened, after applying the quality control of the form and assuring of the right filling of all the required data. The new page will have the results that the model can produce. The new page, which will represent the model simulation under the current weather as a graph, has four tab windows; the first one given in Photo 2 is displaying the daily data of the accumulated above biomass in a graph of the different plant organs, grains, leaves and stem. The daily above biomass dry weight is controlled by two buttons, which are next and previous for displaying the next daily point or cutting the current point and keep only the previous above biomass representative graph lines. Also, by the beginning of each phenological stage or facing the growth any hot episodes or water deficit that will be written below with its dates.

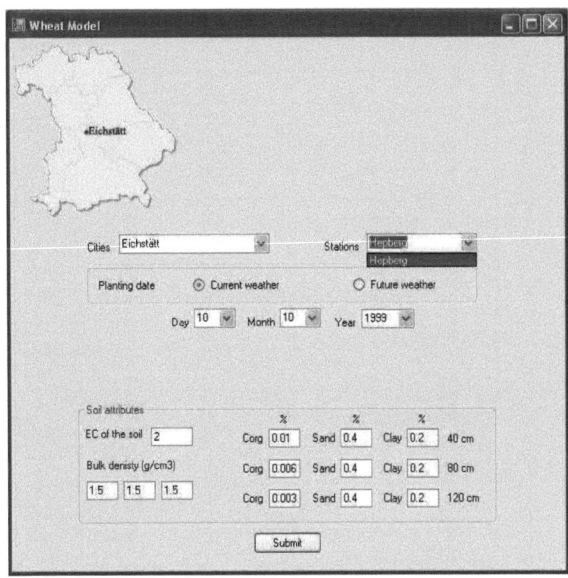

Photo 1. Model form screen displaying selected input variables under current weather options.

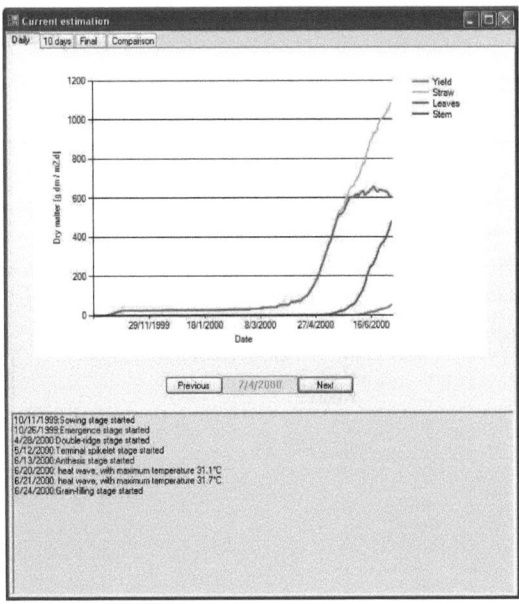

Photo 2. Screenshot displaying daily output values of the growth of different plant organs under current weather conditions.

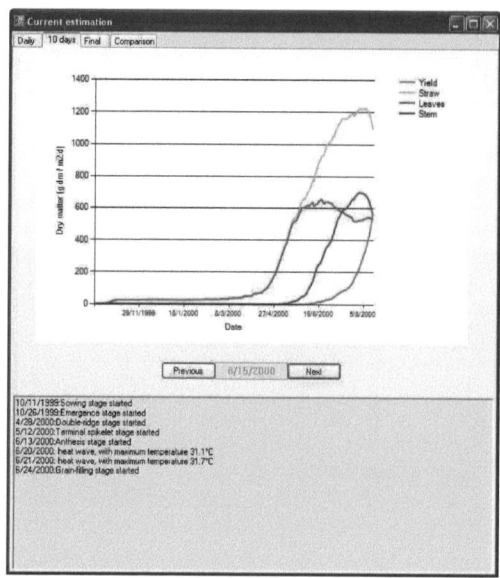

Photo 3. Screenshot displaying 10-days output values of the growth of different plant organs under current weather conditions

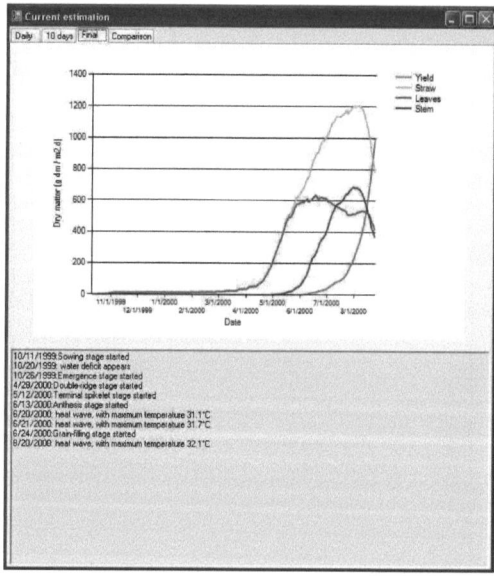

Photo 4. Screenshot displaying the final output of the growth of different plant organs under the current weather data.

The second tab window at Photo 3 is representing exactly as the first one, but with a point each 10 days for a quicker display of the crop performance during the season, with the next and previous buttons also.

The third tab widow at the current estimation of the model at Photo 4 is displaying the complete graph of the winter wheat crop organs performance during the selected season, with writing below the dates of the beginning of all the phenological stages and all the existing water deficits periods and hot episodes, with writing the maximum temperature of that hot episode, during the season. The fourth and last tab window (Photo 5) is displaying an option of making a comparison among the selected season and the other following seasons at the same selected site, the same planting day and month, and soil attributes. The comparison started after selecting a season to represent as the part, which the comparison will end at it, where the comparison will start from the selected season from the first form, and submitting the form by clicking on the compare button. This comparison option clarified the effect of the weather variation on the yield, where by this comparison the weather data will be the only variable factors that affect the growth, the other factors from the location and the soil variables would be the same. The output here will be the final grain yield for the selected seasons represented in a graph, followed by mentioning the number of water deficit periods and the hot episode days, which faced each of the displayed seasons. That last option could be reformed again by selecting a new season from the list box and clicking again on the compare button.

By returning back to the first form, and selecting the future weather option instead of the current weather one, the current weather data will disappear and the future weather available data will be shown, represented in the three studied future periods, where each period is revealed by its last season. As well as, the user can select the planting future weather date for all the studied future models (REMO, CLM and STARII) as in Photo 6 or at only one of them as in Photo 7. After submitting the form, which checks first the right filling of all its fields, a new page will be opened also with four tab windows, for introducing the expected future yield at the selected future weather period. The first and second tab windows are displaying the above biomass (Photo 8) and yield (Photo 9) performance during the thirty seasons of the selected future weather period, by using one or all the weather models. The yield performance tab window is showing also below the expected start of all the phenological stages dates, the occurring of water deficits and hot episodes, provided by the maximum temperatures at these hot episodes, during the thirty seasons of the selected future weather period. The graphs of the above biomass and yield performance for a whole future weather period (2021 – 2050 or 2051 – 2080 or 2071 – 2100) indicated to the expected range of the crop behavior under the projected weather data, not assimilated by year per year, but with the whole period.

The third and fourth tab windows are explaining and comparing the total expected yield of the selected future period, individually as in Photo 10 with all the 30 separated seasons at that future period, or the average yield of the selected future period as in Photo 11. The third tab window is displaying the expected yield variation individually between the single seasons within the selected future weather period under one or for all the weather data sets obtained from the regional climate projections. This comparison clarified the predicted fluctuations of the yield within the same future period. But the comparison of the fourth tab window explained the average yield variation among the different used projections.

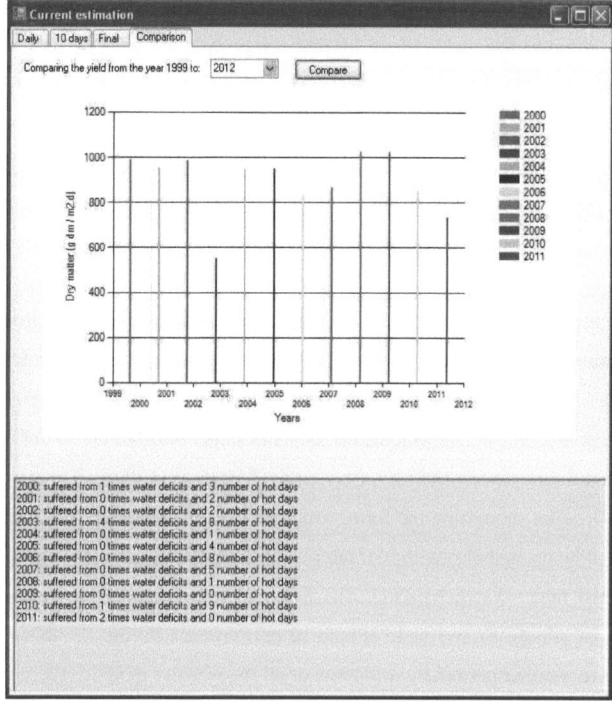

Photo 5. Yield comparison between different seasons under current weather data.

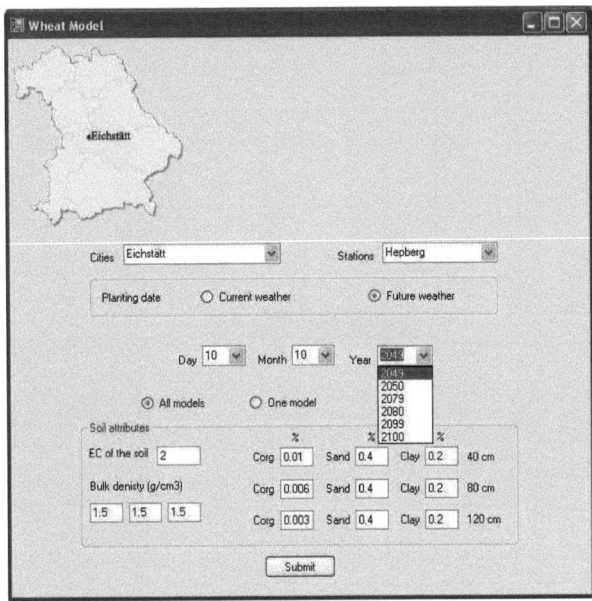

Photo 6. Screenshot displaying selected input variables under future weather options

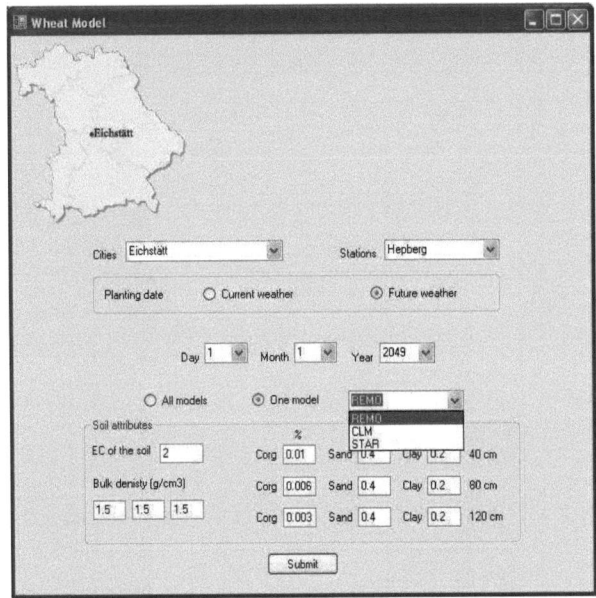

Photo 7. Screenshot with future weather options by selecting the future weather model.

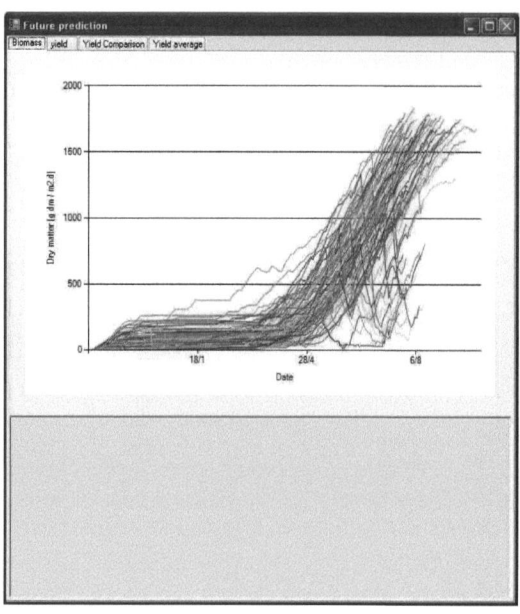

Photo 8. Screenshot of the expected biomass performance during the selected future period.

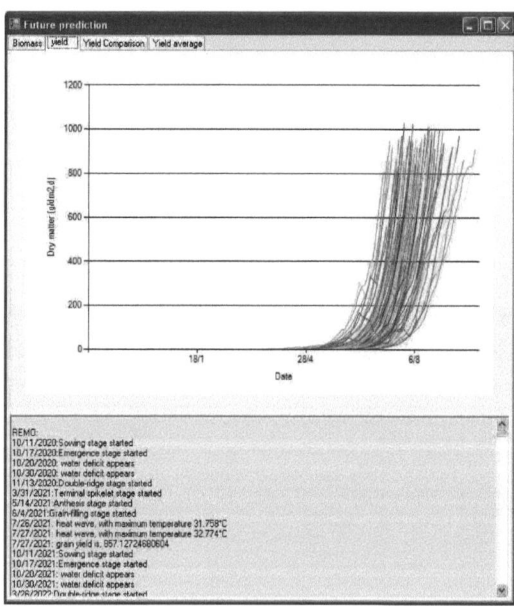

Photo 9. Screenshot of the expected yield performance during the selected future period.

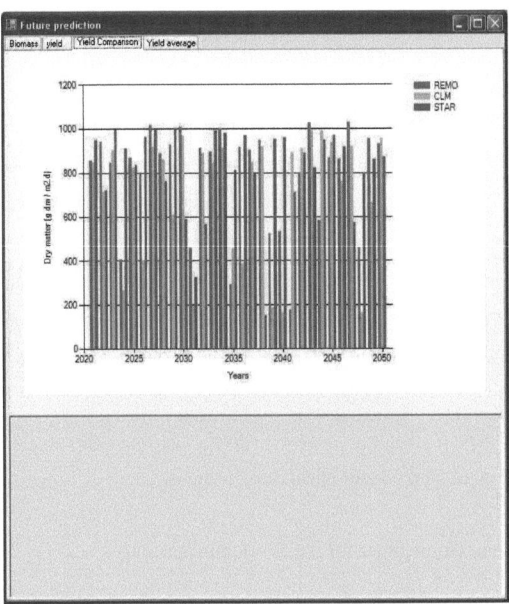

Photo 10. Screenshot displaying expected individual yields at the selected future period.

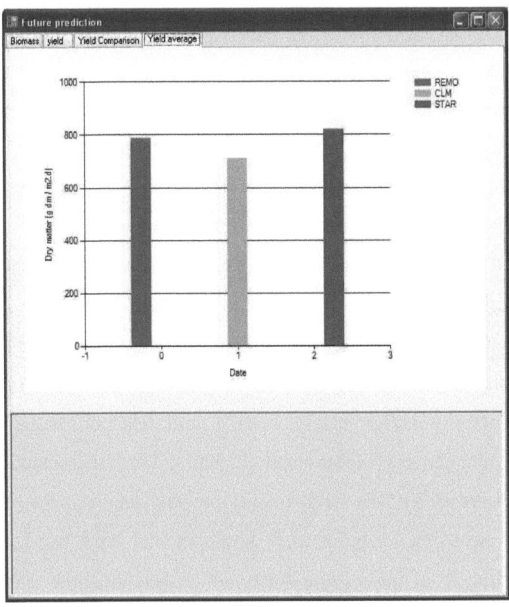

Photo 11. Screenshot of the expected average yield of the selected future weather period.

4 Discussion

Many process-based models, such as crop simulation models or hydrological models require daily site-specific weather data (Jamieson et al., 1998a), and data on soils and management (Brooks et al., 2001) as inputs for estimating the crop behavior according to the surrounding environmental factors, where the farming practices are weather dependent, and yields from single farm vary significantly from year to year depending on each season's weather (Peiris et al., 1996). The developed model here estimates the biomass, yield, and the partitioning of the dry matter into the plant organs during the different developmental stages as shown in fig. 3.

4.1 Temperature and phenological development

4.1.1 Cardinal temperature for the developmental stages

Wheat is generally considered to enjoy an optimum temperature range of 17 – 23°C over the course of an entire growing season, with a T_{min} of 0°C and T_{max} of 37°C, beyond with growth stops (Porter and Gawith, 1999), whilst cultivars seem to differ in their tolerance to extreme temperature (Pomeroy and Fowler, 1973; Blum and Sinmena, 1994; Páldi et al., 1996). The temperature is the main responsible factor for the growth, where temperature is central to how climate influences the growth and development processes of crop plants and is controlled by air or soil temperature (Wheeler et al., 1996b; Batts et al., 1997). The temperature has many typical assumptions, which affect on the crop growth. These typical assumptions are the base temperature (T_{base}), below which development ceases, and a lower optimal temperature (T_{optl}), at which development rate is maximal. The temperature response of the last two temperatures is linear. Maximum development rate is maintained for temperatures from T_{optl} to an upper optimum temperature (T_{optu}). At temperatures above Toptu, the temperature response declines linearly until development again ceases at an upper threshold temperature (Tmax) (McMaster et al., 2008). The last assumptions had been applied on the wheat crop for each of the studied developmental stages on the model. As represented in Table 3, the base temperature for each developmental stage had been set with different temperature, starting from the sowing stage, which had the base temperature 3°C, where Porter and Gawith (1999) had shown that the T_{min} of the period from sowing to emergence is from 2.4 to 4.6°C, T_{opt} from 20.3 to 23.6°C and T_{max} from 31.8 to 33.6°C, and Spilde (1989) mentioned that the suitable temperature range for germination is from 4 to 37°C, which is a wide range. The emergence and double-ridge stages had 5°C as a base temperature, which is slightly higher than other references such as Wilsie (1962) and Petr (1991), who set the base temperature for emergence stage at 3.9 – 4.4°C, Slafer and Savin (1991) set it as 4°C for the emergence and double-ridge phenological stages, and others such as Stapper and Lilly, 2003; Porter and Gawith, 1999; McMaster et al., 2008; Slafer and Savin, 1991 as well, where the used values were more reliable with the model. Slafer and Rawson (1995b) clarified that the spikelets may be initiated at temperatures higher than 1.5°C, therefore the base temperature for the terminal spikelet stage was assumed to be 2°C, and this stage had the lowest temperature range, where the optimum temperatures for this stage lie between 9.3 and 11.9°C, with temperatures greater than 25°C being sub-optimal (Porter and Gawith, 1999). The highest base temperature during the different growth phases was at anthesis, with a temperature of 9.5°C (MacDowell, 1973; Slafer and Savin, 1991; Russell and Wilson, 1994). The minimum temperature values for grain-filling range were from 4.1°C (Hunt et al., 1991) to 12°C for winter wheat (Russell and Wilson, 1994), and because of that wide range, the base temperature for the grain-filling stage was assumed as 7°C. Hence, the consequences of the base temperatures to the different phenological stages agreed with Addae and Pearson (1992) observed a linear relationship between sowing and emergence and during grain-filling and a non-linear relationship between emergence and anthesis. The cardinal temperatures for phenological stages rise steadily with plant development. Thus, base values increased from below zero at germination to above 7°C during grain filling, while optimum values raise from less than 22°C to more than 25°C (Angus et al. 1981; Del Pozzo et al. 1987; Porter et al. 1987; Morrison et al. 1989; Slafer and Savin 1991; Slafer and Rawson 1995d).

4.1.2 Growing degree days for the developmental stages

The base temperature is not the only controlling temperature factor to the crop growth, there are also the thermal degree units (TDU), that control the duration of each phenological stage after gaining the base temperature to get in that stage. Rawson and Macpherson (2000) showed the TDU of the phenological stages sowing, emergence and double-ridge as they have been used here in the model with values of 60, 200, 150°Cd respectively, where Jamieson et al. (2007) mentioned that the photo-vernal-thermal time interval from emergence to floral initiation was increased from 140 to 180°Cd, and the photo-thermal time interval from floral initiation to double-ridge was increased from 130 to 180°Cd, and hence, these last values equal also the used TDU from emergence to double-ridge stages. Terminal spikelet stage was applied in the developed model here as what MacMaster et al. (2008) clarified that the terminal spikelet is reached when thermal development units reaches 400°Cd. And finally the TDU for the grain-filling developmental stage was set with 590°Cd, which is slightly higher than what Semanov (2007) mentioned that 550°Cd are typically required to complete grain filling.

The response of yield is the most irregular for temperature. The main effect of temperature is to set the phenological dates and so changes in temperature affect both the timing and duration of the main growth periods in which most of the biomass is accumulated (Brooks et al., 2001).The calculation of the TDU was used here for setting the different growth stages timing, which is a common method for many simulation models. The major methods have been used in models to simulate the timing of important events in the lifecycle in wheat, and the influence of daylength and vernalisation on their timing. The older of these methods (e.g., Ritchie and Otter, 1985; Weir et al., 1984) describe development through sequential phases (e.g., floral initiation, terminal spikelet, flag leaf appearance) leading to anthesis. Intervals of modified thermal time are assumed to be constant for these phases, and key stages such as anthesis are reached when the sequence of phases has been completed (Weir et al., 1984; Ritchie and Otter, 1985). A variation on this approach was to use thermal time either directly or as represented by the number of leaves between developmental events (McMaster et al., 1992; Jamieson et al., 2007). The AFRCWHEAT2 model (Weir et al., 1984; Porter, 1993) for example was used to simulate wheat, which is a determinate crop, with well-defined development stages, and with a physiological end to the growing season–crop maturity. The timing of these stages is controlled by accumulated temperature, in some cases modified by a combination of photoperiod and vernalisation. The growth rate of the canopy is temperature-driven and limited by water shortage. There are two main reasons for predicting the timing of crop development; first, the effects of environment (temperature and photoperiod) on crop development (phenology) are central to crop adaptation (Evans, 1993; Roberts et al., 1996), second, the accuracy of the phenology sub-model of a crop simulation model is a sensitive step to the precise prediction of crop biomass and yield (Wheeler et al., 2000).

The thermal degree units (TDU) were not only responsible for setting the phenological stages timing, but also for the dry matter accumulation to the crop during these different phenological stages as clarified in the equations 33 to 52. Figure 3 shows the plant dry matter at the growth stages, which was not the same for the same stage at the different seasons in Landshut. For example, the accumulated dry matter of the early stages at the seasons 2003/04, 2004/05, 2005/06, and 2007/08 was very low depending on the existing low weather temperatures at these stages, but it was relatively high at the other seasons specially at the season 2002/03, where the temperature was higher at these stages. The last comparison is taken into consideration only at the early growth stages (sowing and emergence), where winter wheat is most commonly planted in months when the mean daily temperature is between 8 and 16°C (Bunting et al., 1982). This suggests that in practice, winter wheat is generally sown in sub-optimal temperatures and that warmer conditions during sowing should not have a negative impact on wheat establishment (Porter and Gawith, 1999). Nevertheless, the higher temperature of the reproductive stages gave lower dry matter than the lower one, where the reproductive stage are more sensitive to temperature than the vegetative stage, where Halevy (1985) found the period from initiation of terminal spikelet to heading to be more sensitive to temperature than sowing to floret initiation or floret initiation to initiation of terminal spikelet. Therefore, the yields tend to decrease as temperature increases because of the earlier maturity (Brooks et al., 2001). And hence, with the same warm season 2002/03, the harvest date was the earliest, which was at the range from 21/7 at Würzburg to 14/8 at Neumarkt, and the dry matter was the lowest, which was at the range from 38.28 dt/ha at Lichtenfels to 79 dt/ha at Freising, where in the colder season 2003/04, the harvest date was the last, which fluctuated between 16/8 at Würzburg and 27/9 at Neumark, and with a high final dry matter, which fluctuated between 84.3 dt/ha at Würzburg and 99.9 dt/ha at Freising.

Not only the harvest date was different during the different seasons and sites, but also the timing of the developmental stages differed regarding the different weather data, where, for example the sowing stage in Neumarkt for the seasons 2002/03, 2003/04, and 2007/08 with values of 123, 111, and 95 days respectively, which is extremely long, as a result of the very low temperature during these periods to collect the required TDU of that stage. With the same concept also in Freising and Regensburg at the season 2003/04 the sowing stage was 91 and 113 days respectively. With the opposite behavior the double-ridge stage started at some cases considerably earlier than the normal one, where normally it started at April and at the beginning of May, but it was in some cases at January and February. This case appeared at the season 2006/07, where that season at many sites faced a relatively warm winter without a period of negative temperature, and therefore, the accumulation of TDU was quickly resulting in an earlier attainment of the development stages especially at the vegetative stages. The early double-ridge stage appeared in Donau-Ries, Lichtenfels, Main-Spessart, Weissenburg-Gunzhausen, and Würzburg at the dates 24/1, 18/2, 19/1, 19/1, and 10/1, respectively, and at the season 2000/01 in Main-Spessart at 13/1. At the last seasons, the reproductive stages started also earlier, according to the early start of the vegetative stages, but not as early as in the case of the vegetative stages.

4.2 Extreme temperature effect on the growth

The temperature can also have negative effects on plant growth, where both high and low temperatures decrease the rate of the dry matter production and, extremes can cause production to cease (Grace, 1988).

4.2.1 The effect of hot episodes

The plant exposure to high temperature during the growing season affects differently on the growth and yield, depending on the exposed growth stage to the high temperature, where hot temperatures at the time of flowering can reduce the potential number of seeds or grains that subsequently contribute to the crop yield (Wheeler et al., 2000). And the exposure to high temperatures during early spike development reduced the number of kernels per spike (Johnson and Kanemasu, 1983), where the optimum temperature for maximum spikelet number (and indeed for the initiation terminal spikelet) was 15°C and elevated temperatures during tillering increased the number of heads (Porter and Gawith, 1999). The response of the crop during the reproductive stages especially at anthesis and grain filling to the high temperatures was stronger compared to the other growth stages. Therefore, temperatures higher than 31°C and lower than 9°C during the anthesis may therefore be considered as the limits of successful anthesis (MacDowell, 1973; Russell and Wilson, 1994). Plants are most sensitive to high temperatures in the first three days after anthesis (Stone and Nicolas, 1995) during which time grain set may be reduced. However, the number of grains per year at harvest maturity declined rapidly when T_{max} was greater than 31°C during this period. And therefore, high temperatures occurring during grain filling are known to affect grain quality (Blumenthal et al., 1993, 1998; Conory et al., 1994; Stone and Nicolas, 1995; Tester et al., 1995; Sanhewe et al., 1996; Corbellini et al., 1997; Wallwork et al., 1998). The extreme events were simulated by the model at the phenological stages from the terminal spikelet to the grain filling, but the more effective high temperature on the growth was at the anthesis and grain filling stages. For example, the first heat wave faced the season 2002/03 in Landshut as shown in Figure 3, where this season faced 11 hot days, at 6/5/2003, which was at the terminal spikelet stage with maximum temperature 31.1°C, however the decline in the dry matter was small, about 1.5 dt/ha, compared with the deep dry matter decline at anthesis and the beginning of the grain filling stage, which decreased the dry matter nearly by 60 dt/ha, after facing at that period 6 hot days with a temperature range from 31.7 to 36.4, where the last temperature was extremely high for the wheat crop. The last strong decline in the dry matter was enhanced also with a water deficit effect on the plant. The last four hot days in the 2002/03 season declined the total dry matter with about 4 dt/ha. The same behaviour had been repeated with the heat waves at the other seasons, with total dry matter declining as a result of the heat waves, but with less effect than in the season 2002/03 because it faced less extreme hot days. The highest yield at Landshut was at the season 2008/09, which did not face water deficit or heat waves, except for one hot day with 34.2°C, and got an optimal temperature for almost all the phenological stages. Not only Landshut had the extreme temperature effect on the crop growth during the different phenological stages, but also the other studied locations did. In general the season 2002/03 faced the highest heat waves compared to the other seasons, where the summer 2003 was recorded as the hottest in Europe since 1500 (Poumadere et al., 2005). The hot days, which faced the season 2002/03, ranged from 9 to 19 days during all studied locations (Appendix, A1). The severity effect of the hot days depends on the temperature degree, timing, successive hot days, and the existing of the water deficit with the heat wave. The higher the temperature degree, the higher the dry matter reduced. The temperature of 38°C was the highest maximum temperature degree faced by Bavaria, this degree was repeated only twice across the studied locations and seasons, once at Günzburg at the season 2002/03 directly at the beginning of the anthesis growth stage at 12/06 with 38.6°C, and the other was in Weißenburg-Gunzenhausen at the season 2006/07 after the beginning of the grain-filling stage with 10 days at 19/06 with 38.3°C. Both temperatures lasted for only one day, but caused a loss in the total dry matter nearly by 15 dt/ha at that day. The heat waves at the end of the terminal spikelet, the anthesis and the first half of the grain-filling stages were more effective in reducing the dry matter than the second half of the grain-filling stage, even if they faced a longer heat wave. Therefore, the most sensitive period to the high temperature where at anthesis with a few days before and a period from the grain-filling stage, during the grain set duration (Rawson and Macpherson, 2000), where temperatures above 31°C immediately before anthesis reduces grain yield by inducing pollen sterility, thus reducing the grain numbers (Wheeler et al., 1996a). Therefore, it was found that the lowest yield season 2002/03 (Figure 1) got the total dry matter reduced, as a result of the heat waves, during the period from the end of terminal spikelet to the first half of the grain-filling stage. Wheeler et al. (1996b) found that temperatures above 31°C for the five days prior to anthesis resulted in a high number of sterile grains, and hence the exposure to sub- or super-optimal temperatures during anthesis may also reduce yields through the production of infertile florets (Russell and Wilson, 1994). Not only the season 2002/03 faced the last conditions from heat waves, but also the seasons 2004/05 at Lichtenfels, 2005/06 at Passau, 2004/05, 2005/06, and 2007/08 at Main-Spessart, 2004/05 at Weißenburg-Gunzenhausen, and 2005/06 at Würzburg. And hence, the successive hot days during the season had more effect than the one hot day with the same or nearby temperatures, and that appeared at the season 2004/05 in Lichtenfels, where the total dry matter declined during anthesis at 21/06 with 31.4°C maximum temperature and was smaller than at 23-25/06 with temperatures of 31.2, 33.4, and 31.4°C, respectively, and that was also repeated at Passau at the season 2001/02 also during the anthesis, which faced two heat waves, the first one lasting only one day at 15/06 with maximum temperature of 31.7°C, and the second one had four successive hot days 18-21/06 with maximum temperatures of 31.1, 33.8, 32.9, and 31.3°C, where that second heat wave caused a stronger decline in the total dry matter. The appearance of the water deficit with the heat wave resulted in an extreme negative effect on the total dry matter, where in Eichstätt at the season 2002/02 at anthesis, there was a water deficit period starting at 07/06, which was surrounded by hot days at 5, 10, 12, and 14/06 with temperatures of 31.5, 31.6, 34.2, and 34.7°C, respectively. These conditions reduced the total dry matter at that period nearly up to 20 dt/ha.

4.2.2 Freezing effect

Not only the high temperature decreases the rate of the dry matter production, but also the very low one does the same (Grace, 1988). Like many other physiological characteristics thermo-resistance is not a constant property but varies within rather wide limits depending on genotype and external environment (Biebl, 1962; Levitt, 1972). Wheat may be at risk of damage by low temperatures at all stages of crop development. However, susceptibility increases with increasing development of the crop. Conversely, the risk of damaging frosts occurring reduces as spring progresses (Spink et al., 2000). When the wheat crop (Triticum aestivum L.) is in the vegetative state, frost may lead to a decrease in the rate of photosynthesis (Marcellos, 1977) or even to leaf, root and plant death (e.g. Pittman, 1933; Chen et al., 1983; Andrews et al., 1997). As the crop develops, stem extension occurs pushing the apex above the insulating soil surface. At this point, the apex has switched to the production of reproductive primordia (the ear). These reproductive primordia are thought to be more susceptible to frost damage than vegetative primordia, partly due to the more exposed position above the soil and partly due to the nature of the tissue (Single, 1984).

The frost effect had been also applied to the model, starting from the reproductive stages, where wheat is most sensitive to freeze injury during reproductive growth, which begins with pollination during late boot or heading stages. Temperatures that are only slightly below freezing can severely injure wheat at these stages and greatly reduce grain yields (Warrick and Miller, 1999). The negative effect of the frost was applied in the model, if the air temperature decreased below 0°C with the beginning of the double-ridge growth stage, where Mahfoozi et al. (2001a, b) mentioned that the double ridge stage has normally been used as an index of the transition in relation to changes in the level of frost tolerance. Therefore, the frost effect started after the achieving of the vernalization saturation, where then appears that there is a down regulation of the expression of low-temperature tolerance genes (Fowler et al., 1996). Frosts killed spikelets, restricted internode extension (stem growth) and reduced yield. Frosts in April and May, after growth stage (GS) 33, appeared responsible for the damage symptoms observed in the crop (Whaley, et al., 2004).

It was expected that elevated temperature would have a negative effect on frost tolerance, primarily due to less hardening at higher temperatures and less accumulation of carbohydrates (Hanslin and Morensen, 2010). The frost damage appeared to be very low with the simulated seasons and locations in the model, where the simulated crop growth did not face very long and very low freezing temperatures. All the freezing temperatures, which appeared in the reproductive growth stages, were only in the double-ridge stage and did not continue to the terminal spikelet stage. The lowest minimum air temperature appeared during the double-ridge stage in Würzburg at the season 2002/03 from 8-10 April with minimum temperatures of -7.4, -5.4, and -5.8°C, respectively, and hence, with these very low temperatures a slight frost damage happened, and that agreed with Spink et al. (2000), who defined a frost that is likely to damage the crop from the double-ridge stage onwards as when the air temperature falls below -5°C for two consecutive days. The other seasons at the other locations did not face minimum temperature lower than -5°C, but all were higher, with sometimes a longer period in the double-ridge as in Lichtenfels at the season 2006/07, Schweinfurt at the seasons 2000/01 and 2006/07, Weißenburg-Gunzenhausen at the season 2006/07, and Würzburg also at the season 2006/07. The last seasons had earlier double- ridge stages beginning, which were at 18/2/2007, 13/1/2001, 13/1/2007, 19/1/2007, 10/1/2007 respectively with the last seasons, depending on the existing relatively warm winter at those seasons, were the warm winters led to rapid development and less cold hardening, and the occurrence of very cold temperatures in spring resulted in frost damage to the ears (Whaley, et al., 2004). Also the early sowing, which was at season 2000/01 in Schweinfurt, could be a reason of the early beginning of double-ridge and the appearance of the frost damage, where the early sowing of wheat and the use of rapid developing varieties resulted in a serious risk of frost damage (e.g. Single, 1984). These practices give rise to a fast developing crop whose apices are more likely to become reproductive early and therefore susceptible to frost damage (Spink et al., 2000). Nevertheless, the appearance of the freezing temperatures during the double ridge growth stage, but still the effect of the frost damage was very low compared with the high temperature effect on the growth.

4.3 Water deficiency effect on the growth

Not only the high or low temperature had a negative effect on the crop yield, but also the water deficiency to the crop during the growing season. The available water in the soil decreases during the growing season limiting plant growth. Different crops may respond differently to water limitation, depending on their water requirements. Therefore, the effect of drought on crops should be characterized not in terms of the soil water deficit experienced by the crop during the growing season, but by the reduction in grain yield caused by water limitation (Semenov, 2007). Therefore, the effect of the water deficit during the growing season could be recognized from the crop yield or the total dry matter during the season. However, the number of the water deficit periods, which faced the season, and also the timing of it are important on affecting the yield and dry matter. The appearance of the water deficit periods during the sowing and emergence growth stages did not affect strongly the total dry matter production and the yield, since the dry matter at these stages was still small. And hence, the water deficit periods, which appeared mainly at the seasons 2005/06 and 2006/07 for the sites Donau-Ries, Freising, Main-Spessart, Regensburg, and Weißenburg-Gunzhausen, when the weather was relatively dry at the season beginning, while Eichstätt and Günzburg had these earlier water deficit periods only at the season 2006/07, and Lichtenfels and Würzburg at the season 2005/06. Freising faced also an earlier water deficit period at the season 2003/04. Two earlier water deficits periods appeared at the season 2005/06 in Freising at 21, and 23/10/2005 and Lichtenfels at 23/10/2005 and 3/11/2005. These earlier water deficit periods appeared at the second half of October till the first 3 days in November, and these periods had no negative effects on the yield regarding to the small existing dry matter at that time.

The occurrence of the water deficit at the crop growth stages starting from double-ridge till the grain filling showed a clear decline in the total dry matter during the season, depending on how extremely was the water deficit in the soil, how many water deficit periods appeared during the season and if this water deficit period coincided with a heat wave period. Therefore, we can see in Figure 3 that the season 2002/03 at Landshut faced two successive water deficit periods at 12 and 22/6/2003 surrounded by 3 hot days at 10, 12, and 23/6/2003 with maximum temperatures of 32.2, 34, and 36.4°C, respectively, which led to a decrease in the total dry matter at that period with nearly 62 dt/ha. The last case had been repeated in Eichstätt at the season 2002/03 (Appendix, A1), where a water deficit period appeared at 7/6/2003 surrounded with 4 hot days at 5, 10, 12, and 14/6/2003 with maximum temperatures of 31.5, 31.6, 34.2, and 34.7°C, and that decreased the total dry matter with nearly 20 dt/ha. At the same site and season, there were three successive water deficit periods at 29/03/2003, 18, and 28/4/2003, which caused a decrease in the total dry matter with almost 10 dt/ha. The occurrence of many successive water deficit periods during the reproductive growth stages appeared also in Lichtenfels and Main-Spessart at the season 2002/03 with a total dry matter decline of more than 15 dt/ha for both, the water deficit periods for Lichtenfels were at 8, and 18/3/2003, and 8/4/2003, and in Main-Spessart at 18 and 28/3/2003, and 8/4/2003. However, not all the water deficit periods had the same negative effects on the total dry matter, depending on the drought percentage in the soil, where we can find a very small decline in the total dry matter in Weißenburg-Gunzhausen at the season 2003/04, which faced a water deficit period at 22/3/2004. Nevertheless, the appearance of some other single water deficit periods during the growing season caused a significant decline of the total dry matter, as in Lichtenfels and Neumark, where both faced a water deficit period at 2006/07 at 4/5/2007, with WRSI 89.8 and 91.2%, respectively, which reduced the total dry matter with values nearly by 42 and 23 dt/ha, respectively. The severity of the soil drought, which caused the water deficit effect to the plant, depends on the amount of the precipitation or irrigation, the soil properties, and the existing temperature and radiation, where an increase in radiation increases the rate of accumulation of biomass throughout the year and increases the water deficit through greater transpiration by the plant. For low values of solar radiation, the water deficit is small enough that the potential yield is obtained. Both potential anthesis biomass and potential grain fill biomass are proportional to solar radiation and so the yield is also proportional (Brooks et al., 2001).

The soil characteristics are very important in its capacity for holding the water in the soil and an agent for accelerating or resisting the water deficiency in the soil (Minansy and McBratney, 2000). The difference between Figure 3 and 4 is discussing the effect of the soil characteristics on the yield without taken into consideration the weather effect, where both Figure 3 and 4 are displaying the same studied nine seasons in Landshut, which means the same weather data, whereas with using different soil types at each one. The better soil type was at Figure 3, which indicated the observed soil type (silt loam) at the site managed by LfL, while at Figure 4 the used soil type was sandy loam, which had less capacity for holding the water in it than the silt loam type. The seasons 2003/04, and 2006/07 faced water deficit periods with the sandy loam soil, while it did not face it with the silt loam soil, the first season 2003/04 showed four water deficit periods, the first two periods was at the sowing stages and they did not affect significantly on the biomass or yield, but the last two periods were at the reproductive stages and caused a strong decrease in the aboveground biomass nearly by 30 dt/ha after the last water deficit period at the end of May, where the total yield decreased nearly 20 dt/ha compared to the simulated yield at the same season by using the silt loam soil. The second season at 2006/07 faced two water deficit periods, the first one was at the end of the sowing stage, which had also no significant effect on the biomass, and the second was at the end of April and it was a severe water deficit period, which caused a sharp decrease in the aboveground biomass about 35 dt/ha, since the precipitation at previous month (March) was relatively low with value of 29.5 mm, and in April it was totally dry with almost no precipitation (6 mm), which lead to a very dry soil. The total yield at this season was 20 dt/ha lower compared to the total yield simulated using the silt loam soil. The season 2002/03 shown by Figure 3 had one water deficit period during a warm summer with eleven hot days, but at Figure 4 with using the sandy loam soil, the season 2002/03 showed five water deficit periods starting from the end of March till the end of June, in addition to the eleven hot days. The warm summer at the season 2002/03 plus the less water holding capacity of the used sandy loam soil compared to the silt loam one caused quick water soil loss (evaporation), which caused the five water deficit periods, these periods decreased the total yield nearly by 20 dt/ha compared to the same season with the silt loam soil.

For calculating the soil hydraulic properties, which describes the water movement through the soil porous, Pedotransfer functions (PTFs) are becoming increasingly popular for estimating hydraulic properties from data on soil texture, bulk density, organic matter content, and saturated conductivity. The majority of PTFs are completely empirical, although physico-empirical models and fractal theory models have also been developed (Minansy and McBratney, 2000). Particle size distribution is used in almost any pedotransfer function. Different national and international classification systems use often quite different particle size classes. As a consequence, textual classes used in PTFs vary considerably. However, using sand, silt and clay contents is a common approach (i.e. Maclean and Yager, 1972; Pachepsky et al., 1982; Rajkai and Várallyay, 1992; Williams et al., 1992b; Shein et al., 1995; Wösten et al., 1999). The used PTF by the model was that of Vereecken et al. (1989), where Corneils et al. (2001) mentioned that the PTF of Vereecken et al. (1989) was the most accurate. It had the highest ranking for the four validation indices that were computed in this study. It had also a very high applicability to our soils, since it is based on a very wide range of the required soil properties. For calculating the PTF, the parameters wilting point and field capacity need to be used as displayed in equation 23. The wilting point is usually defined as the water content below which plants wilt during the day and cannot recover overnight. Estimated in undistributed soil cores, the 'permanent wilting point' is usually approximated by water content determined at 1500 kPa tension. While the water content at field capacity is defined as the volumetric soil water content after a soil has been thoroughly wetted to saturation and allowed to drain for two or three days (Soil Survey Division Staff, 1993), this definition is physically less stringent and usually determined at 5 (UK), 10 (Netherlands) or 33 kPa tension (US), depending on prevailing hydrological conditions and tradition (Kätterer et al., 2006). And here the US values had been used in the model.

The dry matter allocation into the different plant organs (stem, leaves, roots, grains) has been distributed in the model by using empirical function, depending on Iglesias (2006), which had a detailed real data of the wheat crop during the season for all the plant organs separated on six different experiments, and also depending on the detailed graph of Rawson and Macpherson (2000) in Figure 9, which had the exact timing of each plant organ initiation and growth according to the different growth stages. This empirical function had been used before by other crop simulation models for dry matter allocation into the plant organs. For example, the crop-climate models Sirius-wheat and SURCOS employ a somewhat simpler relation which empirically partitions dry matter to grains (Jamieson et al., 1998a). And hence, the distribution of simulated root dry matter over depth is calculated by an empirical function according to Gerwitz and Page (1974) varying with the thermal time from the sowing date as described by Whitmore and Addiscott (1987). While other models like AFRCWHEAT2 or CERES-Wheat, accumulate dry matter for a period of time prior to anthesis and then convert this dry matter to grain number (Porter and Gawith, 1999).

4.4 The dry matter allocation at different developmental stages

The different plant organs are not all initiated at the same time, but in a specific sequence, which in turn serve the plant requirement for getting the best performance. After the sowing and starting from the germination stage, the produced dry matter in the model was allocated to the roots, in which the roots provide the plant with the water and minerals from the soil under the minimum required water content (Evans et al., 1975). Following to the root growth, the coleoptile, which protects the emergence of the first leaf (Kirby, 1993), appears directly before the initiation of the leaves starts. The start of the leaf emergence appeared at the middle of the sowing stage; to start sharing the roots in the dry matter allocation but still with a small percentage, and hence this percentage increases by the leaf initiation, where the rates of leaf emergence are slower than those of leaf initiation (Kirby, 1985). Therefore, the processes of leaf initiation and development rates increased with the increasing of the mean daily temperature within the optimal temperature range at this stage, and that may be because of increasing the rates and efficiencies of photosynthesis and enzyme activities, which are temperature-dependent (Makan et al., 1987), where by this way the leaf initiation and development rates are almost linear at temperatures between T_{min} and T_{opt} (Kirby, 1985; Baker et al., 1986). The leaf appearance, which was distributed on the internodes, had been activated during the growth stages emergence, double-ridge, and a part from the terminal spikelet (till the beginning of the flag leaf ligule) (Rawson and Macpherson, 2000), after that the dry matter allocation to the leaves had been declined for the initiation of the other organs. Leaf appearance rate is controlled mostly by the temperature of the apical meristem and leaf expansion zones (Jamieson et al., 1995; McMaster et al., 2003). Final leaf number is largely controlled by responses to vernalization and photoperiod (Brooking, 1996; Mahfoozi et al., 2001a; Brooking and Jamieson, 2002). The stem initiation and growth was started directly before the beginning of the terminal spikelet growth stage, with a small value of stem biomass, then this value increased rapidly before the anthesis, where the rate of stem elongation is lower in the vegetative phase than in the reproductive phase (MacDowell, 1973), with rapid stem growth beginning shortly after the terminal spikelet stage of the main shoot apex (Kirby, 1985). The grain set started directly before anthesis and lasts during this stage. The grain set process is very sensitive to high temperatures, especially at the beginning of anthesis, during with the high temperature (above 31°C) at this period, the grain set may be reduced (Stone and Nicolas, 1995) and resulting in a high number of sterile grains (Wheeler et al., 1996b). Then the grain growth started with the grain-filling stage, while the dry matter at this stage is allocated mainly to the grains till it reached to the physiological maturity or the end of the developmental stages (Rawson and Macpherson, 2000). The optimum conditions area for grain growth was found when anthesis occurs during May. Following anthesis radiation levels during the grain filling period are relatively stable while the temperature continues to increase so that the duration of grain filling and, consequently, the total radiation received both decline (Kirby et al., 1998).

4.5 Climate change effect on winter wheat

The model explained the winter wheat growth surrounded by the different environmental factors (temperature, precipitation, radiation, and different soil attributes) at different spatial and temporal patterns in Bavaria at the current situation as shown in Figures 3 and 4 and Appendix A1. But changes to the global climate, notably to regional spatial and temporal temperature patterns (Houghton et al., 1996), from increased atmospheric concentrations of greenhouse gases are predicted to have important consequences for crop production (Parry, 1990). The cardinal temperatures for phenological stages rise steadily with winter wheat plant development. Thus, base values increase from below zero at germination to above 7°C during grain filling, while optimum values rise from less than 22°C to more than 25°C (Angus et al., 1981; Del Pozzo et al., 1987; Porter et al., 1987; Morrison e al., 1989; Slafer and Savin, 1991; Slafer and Rawson, 1995d). But the global production of annual crops will be affected by increases in mean temperatures of 2-4°C expected towards the end of the 21st century. Within temperate regions, current cultivars of determinate annual crops will mature earlier, and hence yields will decline in response to warmer temperatures (Wheeler e al., 2000).

4.5.1 The used weather climate models

For predicting the future yield for the winter wheat crop under the next 100 years, the future weather data was needed first for this estimation. General Circulation Models (GCMs) are tools designed to simulate time series of climate variables globally, accounting for effects of greenhouse gases (GHGs) in the atmosphere (Prudhomme et al., 2003). They are good for the prediction of large-scale circulation patterns (Bardossy et al., 1997). Methodologies to model the hydrologic variables (e.g. precipitation) at a smaller scale based on large-scale GCM outputs are known as downscaling. They include dynamic downscaling, which uses complex algorithms at a fine grid-scale describing atmospheric process nested within the GCM outputs (Jones et al., 1995), commonly known as Limited Area Models or Regional Climate Models (RCM), such as the used models REMO, which was run on 10 km horizontal resolution (0.088 degree), and CLM, which was run onto a non-rotated grid with 0.2 degree spatial resolution, by using the A1B scenario, which describes a possible future world of very rapid economic growth, global population peaking in mid-century and rapid introduction of new and more efficient technologies with a balance across all energy sources, and statistical down-scaling, as the STARII model, that produces future scenarios based on statistical relationship between large-scale climate features and hydrologic variables like precipitation (Wilby et al., 1998; Goldstein et al., 2004).

4.5.2 The predicted yield with different regional climate models

The predicted future yield at the first future period in Figure 5 showed a relatively high yield at the three studied future weather models, with the highest yield range in general for STARII then REMO and the lowest one was at CLM in most of the cases, where the crop model under the STARII data predicted the lowest number of extreme hot days during the seasons and this is the disadvantage of the statistical models. The predicted extremely hot days at STARII during the first future period starting from 16 days at Günzburg with an average of 0.5 hot days per year, till 371 hot days at Main-Spessart with an average of 12.4 days per year. The model with STARII data predicted also other low hot days seasons at Landshut, Regensburg, and Neumarkt with hot days of 42, 69, and 55 days, respectively, with an average for each of 1.4, 2.3, and 1.8 days per year, respectively. While the model with REMO data predicted for the first future period a closer range of heat waves with STARII but

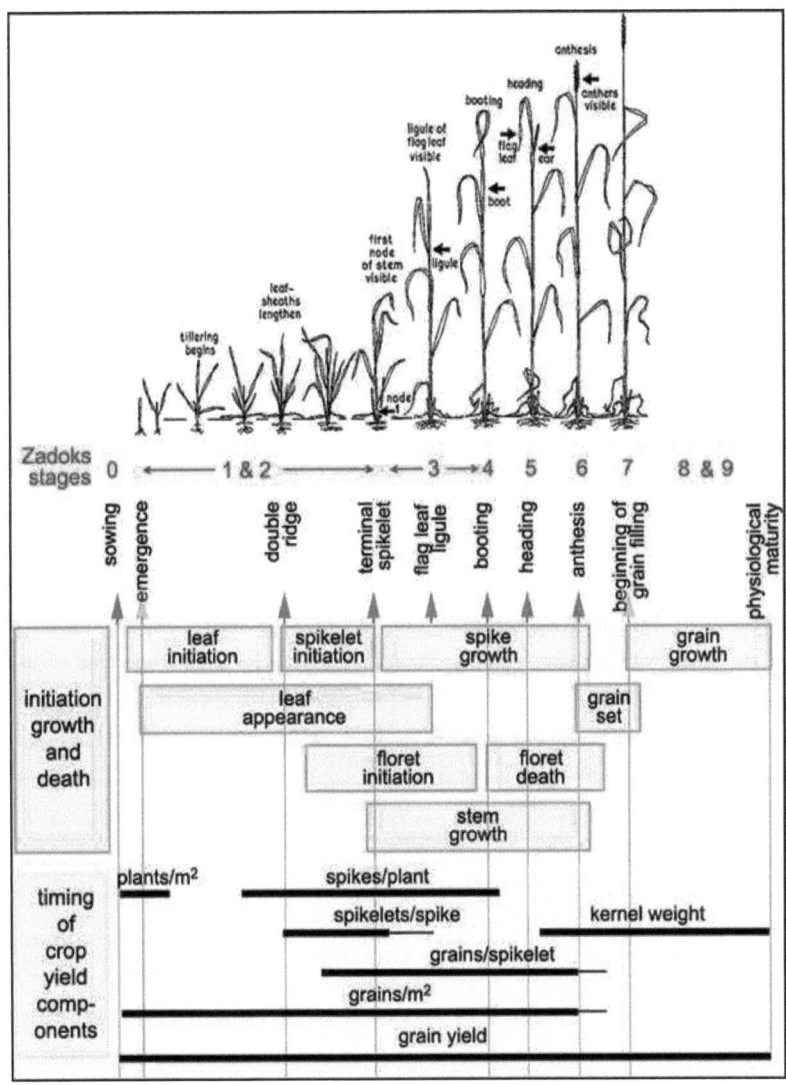

Figure 9: The timing of the initiation growth and death of the different plant organs during the different crop growth (Zadoks) stages in winter wheat (Rawson and Macpherson, 2000).

higher one from 117 hot days at Neumarkt with an average of 3.9 days per year to 274 hot days at Passau with an average of 9.1 hot days per year. The highest heat waves was predicted under the CLM data at the first future period with a range from 185 hot days at Lichtenfels with an average of 6.2 hot days per year to 501 hot days at Würzburg with an average of 16.7 hot days per year. Therefore, the crop model with the CLM data had the lowest yield compared with the other used models' data, and that agreed with Wollenweber et al. (2003), who mentioned that increasing climatic variability and more frequent episodes of extreme conditions may result in crops being exposed to more than one extreme temperature event in a single growing season and could decrease crop yields to the same extent as changes in mean temperature. The crop model yield at that period for all the studied weather models fluctuated between 80 – 100 dt/ha for all the studied sites, except for Main-Spessart, Weissenburg-Gunzenhausen, and Würzburg, which had a yield range between 60 – 100 dt/ha.

The predicted total yield by using REMO and CLM at the second and third future periods showed a lower yield than for the first future period with ranges of 25.6 – 110 dt/ha and 56.14 – 177.9 dt/ha, respectively, for REMO and with ranges of 74.7 – 162.0 dt/ha and 106 – 198.1 dt/ha, respectively, for CLM. That yield declined relates better to the increase of the mean maximum and minimum temperatures of the second and third future periods compared to the first one (Figure 6), where Wheeler et al. (2000) found that increasing the mean seasonal temperature by 2°C will decrease the grain yield by 7%, and the rapid decline in grain yields was associated with a reduction in the number of grains per year at the time of harvest maturity. With the same concept the third future yield was lower than the second one, regarding to the higher mean temperature of the third period compared to the second one, except at Würzburg, where the model showed a higher yield by REMO at the third future period compared to the second one with values of 17.7 dt/ha, although at the same site for the CLM data, the predicted future yield for the third future period was lower than the second with 36.4 dt/ha difference, and that related to the relatively higher hot episodes in Würzburg by the REMO at the second future period than the third one, where the highest predicted maximum temperature that faced the second future period was at 10/8/2037 with maximum temperature of 41.3°C, and it was in the third future period at 9/8/2086 with maximum temperature of 40.2°C, nevertheless the total extreme hot days that had been predicted to face the second future period was 488 hot days with an average 16.3 hot days per year, while the total extreme hot days, which had been predicted at the third future period was 583 hot days with an average of 19.4 hot days per year.

4.5.3 The expected temperature increase effect on the growth

It is clear that changes to the variability of temperature, separate to changes in mean seasonal temperature, affect the yield of annual crops. The effects of brief episodes of hot temperatures on the number of yield components can be particularly dramatic. However, the impact of crop yield cannot simply be predicted from the absolute temperature. Instead, it is reflected by the combination of the magnitude and duration of the hot temperature episode, and coincidence with the development stage of the crop (Wheeler et al., 2000). The episodes of hot temperatures increased at the second and third future periods significantly compared to the first, where the hot episodes by using REMO data for the second future period increased to be in a range from 364 to 692 hot days, while for the third future period they were in a range from 420 to 846 hot days. By using CLM data, the hot episodes were more dramatically, where the predicted range of the second period ranged from 400 to 817 hot days, and from 514 to 966 hot days for the third future period, with highest maximum temperature reached to 46.5°C at 4/7/2094 in Würzburg. If these temperatures increase as predicted, then wheat can mature much earlier compared with the current climate. This has the following consequences. First, the duration of the grain filling stage, measured in calendar days, decreases resulting in the lower amount of radiation intercepted by the plant during grain filling. This, in turn, reduces production of new biomass during grain filling decreasing the final yield. Secondly, the grain filling stage will occur early in the season, when expected daily radiation is sub-optimal, i.e. lower, on average. This reduces grain yield even further. As a result, wheat yield decreases with global warming, if other factors are taken out of consideration (Semenov, 2007).

The season 2061/62 shown in Figure 6 indicated almost no yield (0 – 5 dt/ha) at the sites Donau-Ries, Eichstätt, and Würzburg by CLM future weather data, where this season was predicted to face 45 hot days with a range between 31 to 37.7°C, 45 hot days with a range between 31.2 to 37.92°C, and 67 hot days with a range between 31.1 to 41.3°C, respectively. The same case was repeated also by CLM data at the season 2078/79 in Freising, Günzburg, Main-Spessart, and Würzburg with 30 days (from 31.1 – 40.9°C), 37 hot days (from 31.1 – 42°C), 33 hot days (from 31.4 – 42.9°C), 24 hot days (from 31.4 – 42.9°C), and 36 hot days (from 31.2 – 44°C), respectively. Also the predicted season 2074/75 faced an almost zero yield at Main-Spessart under REMO data with 32 hot days (from 31.3 – 37.4°C) and at Regensburg under CLM data with 30 hot days (from 31 – 41.6°C). With all those hot episodes the yield will be dramatically declined, where the seed yield is particularly sensitive to brief episodes of hot temperatures (Wheeler et al., 2000), and beyond the Tmin and Tmax of the winter wheat, which are 0 and 37°C, respectively, the growth stops (Porter and Gawith, 1999).

4.5.4 The effect of precipitation decline on the growth

Not only the increased temperature and hot episodes affected the predicted future low yield, but also the predicted decrease of the precipitations during the future periods, especially in the summer season impacts on the yield, since changes in the precipitation affect the amount of water in the soil and hence leads to yield loss due to the water deficit (Brooks et al., 2001). The decreasing of the available water in the soil during the growing season limits the plant growth (Semenov, 2007). Therefore, at high precipitation levels, there is sufficient water for the crop and so changes in the precipitation have no effect. As precipitation reduces, both the length and severity of the water deficit increases and so each reduction in precipitation causes a progressive drop in yield (Brooks et al., 2001).

4.5.5 The direct CO_2 effect

Kersebaum et al. (2009) discussed the effects of climate change on wheat production and management across Germany that it will vary depending on the specific regional projection and site properties. Without considering the effect of elevated CO_2 the crop yield under climate change was reduced compared to the reference time slice. Only at one higher elevated site crop growth benefited from increasing temperature. However, including the direct effect of elevated CO_2 on photosynthesis and the indirect effect of reduced transpiration in the simulation yielded in an increase in crop yields in many cases. Only the regions of East-Germany may suffer from more frequently occurring dry spells, where the CO_2 effect was estimated to be not sufficient to level out the negative impact of water shortage on crop growth.

Along with changes in temperatures and shifts in precipitation pattern, periods of drought are predicted to increase in the future (IPCC, 2001). While large uncertainties remain about the extent of these climate changes at the regional scale, increasing atmospheric concentration of carbon dioxide (CO_2) is among the most predictable aspects of global environmental change (Burkart et al., 2004). Historical and modern records show that the atmospheric carbon dioxide (CO_2) concentration increased from approximately 280 μmol mol−1 in pre-industrial times to about 315 mmol mol−1 by 1958, and to more than 350 mmol mol−1 by 1988 (Boden et al., 1994). The accelerated trend in the global CO_2 growth rate during the first 30 years of modern records has led to various scenarios for the future CO_2 concentrations of the atmosphere (Hunsaker et al., 2000).

The direct CO_2 effect on the plant is not yet applied here at the developed model, according to the limited understanding of the combined of climate and direct CO_2 effect on the whole crop growth duration and organs at the different water and fertilizers applications, nevertheless, that it is known that the higher atmospheric CO_2 concentrations lead to higher CO_2 concentration within the plant cells and increases photorespiration. Therefore, some of the CO_2 effect suggestions on the plant by other references will be discussed as follows:

Prichard and Rogers (2000) found that CO_2 had an effect on shoot:root ratio and root biomass. Elevated CO_2 resulted in increased biomass allocation to roots and had earlier been found to stimulate total root length and root biomass, and to produce an increase in root nonstructural carbohydrates. By a closer look at the changes in leaf and root fractions of the total biomass in response to CO_2 as discussed by Leonardos et al. (2003), who showed that in response to elevated CO_2 the leaf fraction decreased and root fraction increased. The drop in soil temperature during autumn usually lags behind the drop in air temperature, allowing roots to continue growing for longer periods than above-ground organs (Hanslin and Morensen, 2010).

Morison (1998) showed that elevated CO_2 causes partial stomatal closure thereby decreasing leaf transpiration at the same time that carbon assimilation is increased. Therefore, because soil water availability is a major environmental factor for plant growth, it is of particular importance to analyse possible interactions of elevated CO_2 and water supply in terms of the water use of plant canopies (Burkart et al., 2004). Nevertheless, Burkart et al. (2004) had also concluded that when water supply is adequate (i.e. high PAW) elevated CO_2 can reduce transpiration. Under drought stress (i.e. low PAW), when soil water content is the main determinant of canopy water use, elevated CO_2 no longer reduces canopy conductance and hence water use. Under drought stress conditions the stimulating effect of high CO_2 on root growth may even enhance canopy water use.

On the other hand Kruijt et al. (2008) found that rising atmospheric carbon dioxide contributes to global warming and thus to changes in both precipitation and evapotranspiration (ET). CO_2, however, also directly affects the productivity and functioning of plants, stimulating biomass growth while reducing the associated water use through reduced stomatal conductance. Their study suggested that direct effects of CO_2 reducing evapotranspiration can be expected to be moderate, up to 5% in the coming 50 years and up to 15% by 2100, with relatively stronger effects in summer and in rougher, natural vegetation such heath lands and (deciduous) forests, where reduced ET leads to less depletion of soil water, hence less water stress on ET and growth and thus less reduced (or more sustained) ET. This is an important feedback effect, especially in drier climates and with natural vegetation, where harvesting does not limit the growth period.

But only Mitchell et al. (2001) were able to show that elevated CO_2 did not affect the relationship between crop evapotranspiration (EC) and plant available soil water content (PAW) of wheat plants grown under controlled environment conditions over a wide range of PAW. The situation is even more complex under field conditions where root growth is known to be stimulated by either elevated CO_2 or water deficit (Katterer et al., 1993; Wechsung et al., 1999) allowing plants to exploit additional water resources. At last, the responses of canopy evapotranspiration (EC) to elevated CO_2 have been found to vary from positive (Chaudhuri et al., 1990; Kimball et al., 1994; Hui et al., 2001), to unchanged (Jones et al., 1985; Hileman et al., 1994; Ellsworth, 1999) to negative (Ham et al., 1995; Drake et al., 1997; Kimball et al., 1999). The variability of these responses may result from several feedback mechanisms that effect EC, including irradiance, wind speed, canopy temperature, VPD, leaf conductance, canopy leaf area (Kimball et al., 1994; Wilson et al., 1999).

Tubiello et al. (1995) found that rainfed crops were found to be more sensitive to CO_2 increases than irrigated ones. On the other hand, low nitrogen applications depressed the ability of the wheat crop to respond positively to CO_2 increases. In general, the positive effects of high CO_2 on grain yield were found to be almost completely counterbalanced by the negative effects of high temperatures. Depending on how temperature minima and maxima were increased, yield changes averaged across management practices ranged from -4% to 8%.

It is in fact well-recognized that CO_2 concentration and management factors will interact in complex ways to determine the ultimate impacts of climate change on crop production. While elevated CO_2 alone tends to increase growth and yield of most agricultural plants (Kimball, 1983; Cure and Acock, 1986; Allen et al., 1997; Kimball et al., 2002), warmer temperatures and changed precipitation regimes may either benefit or damage agricultural systems (e.g., Rosenzweig and Hillel, 1998). Water and fertilizer application regimes will further modify crop responses to elevated CO_2 (e.g., Reilly et al., 2001). Understanding of the combined effects of climate and CO_2 concentration on crop growth and yield is still limited (Ewert,

2004) and application of the present generation of crop models to estimate actual yields for regional and larger scales is critical (Ewert et al., 2002; Tubiello and Ewert, 2002).

4.6 Future expected yield spatial distribution in Bavaria

The spatial distribution of the estimated yield in Bavaria for all the used future periods of REMO and CLM models revealed a strong relationship between the estimated yield, grain-filling duration and the temperature, as shown in the Maps 2 – 7, where with the increase of the number of hot episodes during the studied future periods, the yield and the grain-filling duration decreased. Thus at the same future period, the regions with high occurrence of hot episodes showed lower yield than the other parts of Bavaria. The estimated yield of winter wheat was relatively low at Unterfranken, the north half of Schwaben and Oberbayern, and the southwest part of Niederbayern with a range of 30 – 75 dt/ha at the three future periods for CLM (Map 2), where the occurrence of the hot days were high from 10 to 22 days at the three future periods (Map 3), and that increase of the hot episodes occurrence decreased also the grain-filling duration to be within the range of 75 – 200 days within the studied future periods (Map 4), according to the quick gathering of the thermal degree units of that stage. Whereas the estimated yield was high at Oberfranken, the eastern part of Oberpfalz and Niederbayern, and the south half of Schwaben and Oberbayern with a range of 80 – 92 dt/ha, with lower occurrence of the hot episeodes during the season with a range of 0 – 6 days, resulting in a long grain-filling duration with a range of 95 – 250 days at the three future periods for CLM. The same behavior indicated by using REMO data, where the low yield regions in Bavaria of the three periods were the same regions with high appearance of hot days and lower grain-filling duration, and vice versa with the high yield regions.

With the increase of the predicted average temperature and the existence of hot days for REMO and CLM models gradually within the three studied future periods, the crops were exposed to more than one extreme temperature event in a single growing season, which could decrease crop yields to the same extent as the increase in the mean season temperature (Wollenweber et al., 2003). Therefore, the third period was estimated by the developed model to have the more dramatic effect with the lowest expected yield and grain-filling duration, taking into consideration the same variation among the different regions in Bavaria. However, the estimated yield at the first future period was within the range of 57 – 90 and 87.7 – 96 dt/ha, suffering of hot days within the range of 0 – 17.5 and 0   5.5 days for CLM and REMO, respectively, whereas the last range of yield decreased at the second and third periods to 41 – 89.6 and 42.6 – 85.8 dt/ha for CLM, respectively, and 73.7 – 94 and 76.6 – 91 dt/ha for REMO, respectively, where the indication of the extreme maximum temperature increased compared to the first period to be 0 – 22 and 1 – 19 days for CLM, respectively, and 0 – 7.4 and 0.5 – 6.5 days for REMO, respectively. It has been noticed that the maximum appearance of the extreme hot episodes at the second period of REMO and CLM were higher than at the third period of REMO and CLM (Maps 2, 3, 5 and 6), nevertheless the minimum appearance of the hot episodesat the third period of REMO and CLM were higher than at the second period and also the total expected average monthly temperature of the third future period was higher than the second one (Figures 7 and Appendixes A2 and A3), where the effects of hot temperature episodes are relatively close to the effect of the whole increase in mean seasonal temperature of about +2°C (Wheeler et al., 2000).

The estimated lower yield areas at the studied future periods in Bavaria were located mainly at Unterfranken and the north of Schwaben and Oberbayern, where the hot temperature episodes were higher than for the other regions, and the estimated higher yield areas were located at Niederfranken and the east and south of Bavaria. The estimated yield of Würzburg and Main-Spessart, which repreasented Unterfranken, in Figure 5 was also low at the three future periods with values of 68, 50, 46, 70, 61, and 62 dt/ha for Würzbug and 73, 60, 55, 68, 60 and 50 dt/ha for Main-Spessart, respectively, for CLM and REMO models, regarding to the high estimated hot episodes at Würzburg and Main-Spessart, which were at the range of 7 – 19.4 days at REMO and at the range of 11 – 26.7 days at CLM. As well as, the lower appearance of the hot episodes during the three studied periods was found at Lichtenfels, which represented Niederfranken, and Neumarkt, which represented the east of Oberpfalz, with a range of 3.9 – 14 and 3.9 – 17.2 daysm, respectively, for REMO and 6.2 – 17.1 and 7.8 – 20 days, respectively, for CLM, whilst the estimated yield of Lichtenfels was the highest for the used weather models, which had values at the three future periods of 90, 81, and 80 dt/ha, respectively, for REMO, and 87, 75, and 70 dt/ha, respectively, for CLM. The estimated yield for the CLM data was relatively low at the middle of Bavaria, and relatively high at the south of Bavaria, but fluctuated from high to low for REMO at the same regions inversely proportional with the occurrence of the hot episodes.

The southernmost of Bavaria is all mountainous with very high altitudes (more than 1000 m), therefore, the expected temperature at the three studied future periods for CLM and REMO models gave very low temperatures at these mountainous regions for the first future periods and increased slightly at the second and third one. And as a result of the high altitudes and very low temperatures during the whole season at that part of Bavaria, the grain-filling duration turned out to be very long compared to the other parts of Bavaria, where the estimated grain-filling duration for the REMO at the far south of Bavaria was 230 – 415, 160 – 320, and 125 – 214 days at the three future periods, respectively, whilst the other parts of Bavaria had a significant lower duration than the last one. That difference in the estimated grain-filling duration between the two parts of Bavaria reached to 50 – 320 days at the first future period, which decreased slightly at the second future period to be 45 – 230 days and decreased again at the third future period to be 10 – 115 days, according to the general increase of the temperature at the last future period compared to the first two models as shown in Figure 7 as an example for some locations in Bavaria. The same happened with the estimated grain-filling duration with the CLM data, where the difference of the grain-filling duration between the far south of Bavaria and the other parts were within the ranges of nearby 50 – 300 days for the first two future periods, and within lower ranges at the third future period with ranges of 40 – 220 days. The mountain area of the far south of Bavaria is not a suitable area for winter wheat growing, but the purpose of the spatial distribution of the modeled winter wheat yield is to develop an estimated yield map of Bavaria for distributing the expected winter wheat yield at the future depending on the expected weather data and the existing soil attributes across Bavaria.

5 Summary

A new simulation model was developed as a computer-based model, which calculates the yield and biomass formation for winter wheat from the accumulated thermal units and simple partitioning rules. The presented new crop simulation model, at which principles it was developed and its calibration and validation steps used data from independent experiments from various locations in Bavaria, Germany. The temperature was the main affecting factor on the growth by calculating the base temperature and the accumulated thermal units, which is the most common methodology used by other plant models. The accumulated thermal units characterized the plant output like yield and biomass. The accumulated thermal units in this model were derived from the daily average temperature for partitioning the different growth developmental stages and the allocation of the dry matter into these stages for the different plant organs. The other weather data like solar radiation, precipitation, and relative humidity (the effect of N is not yet included) had an important role in calculating the water requirements of the crop, which also differ according to the developmental stage, and the soil water capacity depending on the soil type. Different environmental factors such as weather factors (maximum and minimum temperature, precipitation, solar radiation, relative humidity, and wind speed) and soil attributes (sand, clay, loam, and organic carbon percent) result in marked differences of winter wheat biomass and yield at the different temporal and spatial scales in Bavaria.

The goal of such models was to simulate and explain crop development and behavior, yield and quality as a function of environmental and management conditions or of genetic variation. The crop simulation model was designed as a simple process oriented model, which computes canopy-level dynamics directly, using empirical relationships without consideration of underlying processes, typically using daily time steps. The purpose of developing a simple process-oriented model was, first to avoid unrealistic results that could result from using very simple statistical models and also to avoid the complexity of the complex process oriented models, where the developed crop model is aimed to be a simple model with available input data for all types of users, whereas it retains the important aspects of the crop growth operations.

The development of the crop simulation model was followed by the calibration and validation processes at twelve sites in Bavaria, which covered almost all Bavaria, with nine crop seasons for each site. The model efficiency revealed an acceptable performance compared to the observed data. The computer-based crop model stores all the weather data for the selected twelve sites in Bavaria for almost 20 years (1990 – 2011) at the model database, where the user will select only the planting date and then the model will grab the weather data for the simulated crop season starting from the selected planting date, till the simulated harvest date. Nevertheless, the user will have to indicate the soil attributes (sand, clay, loam, and organic carbon percent) at different depths. The model will display the crop performance including the simple partitioning rules during the crop growing season and the total yield as the output, offering a yield comparison between different seasons.

The developed crop model has been run under the expected weather for the next 100 years for predicting the expected future yield. The future weather data are represented in three periods (2021 – 2050, 2051 – 2080, and 2071 – 2100) for evaluating the effect of climate change on the total yield. The future weather data were taken from three different models. The first model is STARII, which is a statistical model, and its data covered only the first studied future period. The other two models are the Regional Climate Models (RCM) CLM and REMO, and they covered the three studied future periods.

In general, the simulated yields of the crop models indicated relatively high yields at the first future period compared to the second and third future periods, regarding to the total increase of the maximum and minimum temperatures of these periods, in addition to the increase of the hot episodes appearance during the crop season at the second future period and even more at the last future period. The highest estimated yields at the first future period appeared by using the STARII model compared to the regional climate models, where the STARII model demonstated the smallest appearance of hot episodes. The CLM model at the three studied future periods at almost all the studied sites revealed the highest number and maximum temperature of the expected hot episodes, therefore, the crop model showed always the lowest expected yield with the CLM model compared to the other models.

The purpose of this study was to develop a new crop simulation model that simulated the crop biomass and yield of winter wheat from accumulated thermal units and simple partitioning rules under different soil water and weather conditions in Bavaria, by using the least required input data from the users, which permits a large fraction of users to run the model. Therefore, the simulated crop model was designed to be used not only at the research sector, but also at the educational and applied sectors.

6 Zusammenfassung

Ein neues Simulationsmodell wurde als Computer basiertes Modell entwickelt, welches Ertrag und Biomasse von Winterweizen aus Temperatursummen und einfachen Partitionsregeln berechnet. Die Entwicklung dieser Regeln sowie deren Kalibrierung und Validierung erfolgte anhand von Daten aus verschiedenen, unabhängigen Feldexperimenten auf unterschiedlichen Standorten in Bayern, Deutschland. Die Temperatur war der maßgebliche Einflussfaktor des Pflanzenwachstums im Modell, indem von einer Basistemperatur ausgehend über die Temperatursumme die Übergänge zwischen den einzelnen Phasen des Wachstums definiert wurden. Diese Methodik ist die in Pflanzenwachstumsmodellen am weitesten verbreitete. Die Temperatursumme bestimmte die pflanzlichen Erträge, wie z.b. den Kornertrag und die Biomasse. Die Temperatursumme in diesem Modell setzte sich aus den Tagesmitteltemperaturen zusammen und diente der Unterteilung der einzelnen Wachstumsphasen sowie der Zuordnung der Trockenmasse zu den Pflanzenorganen in den unterschiedlichen Phasen. Die weiteren klimatischen Daten wie die Sonneneinstrahlung, der Niederschlag und die relative Feuchtigkeit (der Effekt des Stickstoffs ist noch nicht berücksichtigt) waren von herausragender Bedeutung in der Berechnung der Wasserbedürfnisse von Winterweizen, welche sich je nach Entwicklungsstadium und Bodenwasserkapazität abhängig vom Bodentyp unterschieden. Die Unterschiede in den klimatischen Bedingungen (maximale und minimale Temperaturen, Niederschlag, Sonneneinstrahlung, relative Feuchtigkeit) und in den Bodeneigenschaften (Textur und Gehalt an organischem Kohlenstoff) resultierten in deutlichen zeitlichen als auch räumlichen Unterschieden in der Biomasse und im Ertrag von Winterweizen in Bayern.

Ziel solcher Modelle ist es, die Entwicklung und das Verhalten von Winterweizen hinsichtlich dessen Ertrags- und Qualitätsbildung durch eine Funktion aus Umwelt- und Managementbedingungen sowie genetischer Variation zu simulieren und zu erklären. Das hier vorgestellte Simulationsmodell wurde als einfaches prozessorientiertes Modell konzipiert, welches üblicherweise in einer täglichen Auflösung die Dynamik der Bestandesentwicklung direkt berechnet, indem es empirische Beziehungen verwendet ohne zugrundeliegende Prozesse zu berücksichtigen. Die Entscheidung für ein einfaches prozessorientiertes Modell wurde getroffen, um mögliche unrealistische Ergebnisse eines einfacheren statistischen Modells oder die Komplexität eines komplexeren prozessorientierten Modells zu vermeiden. Das Pflanzenmodell sollte als einfaches Modell konzipiert werden, welches mit allgemein verfügbaren Daten gespeist werden kann und gleichzeitig die bedeutsamen Aspekte der Wachstumsmodellierung beinhaltet.

Die Entwicklung des Pflanzenwachstumsmodell wurde gefolgt von einem Prozess der Kalibration und Validierung auf der Grundlage von Feldversuchen, die über neun Vegetationsperioden hinweg an zwölf Versuchsstandorten durchgeführt wurden, welche gleichmäßig über ganz Bayern verteilt lagen. Das Modell zeigte annehmbare Ergebnisse verglichen mit den Beobachtungswerten. Das Pflanzenwachstumsmodell beinhaltet in seiner Modelldatenbank alle klimatischen Daten für diese zwölf Standorte in Bayern über einen Zeitraum von mehr als 20 Jahren (1990 - 2011). Aus diesen muss der Nutzer nur den Saatzeitpunkt auswählen, damit das Modell ausgehend von diesem Saatzeitpunkt die klimatischen Daten für die simulierte Vegetationsperiode bis hin zu einem simulierten Erntezeitpunkt festlegt. Darüber hinaus muss der Benutzer die Bodeneigenschaften (Textur und Gehalte an organischem Kohlenstoff) in den unterschiedlichen Bodentiefen angeben. Das Modell stellt dann die Entwicklung der Gesamtpflanze und der unterschiedlichen Pflanzenteile über die Vegetationsperiode hinweg dar, gibt den Kornertrag als Endergebnis an und ermöglicht einen Vergleich der Erträge zwischen den Vegetationsjahren.

Das Pflanzenwachstumsmodell wurde mit den prognostizierten Wetterdaten der kommenden 100 Jahre gespeist, um zukünftige Ertragsentwicklungen zu prognostizieren. Die zukünftigen klimatischen Daten wurden in drei Perioden unterteilt (2021 - 2050, 2051 - 2080 und 2071 - 2100), um den Effekt des Klimawandels auf den Gesamtertrag bestimmen zu können. Diese Wetterdaten stammten aus drei verschiedenen Prognosemodellen. Das erste Modell war STARII, welches ein statistisches Modell ist und dessen Daten nur die erste Periode abdeckten. Die anderen beiden Modelle waren die regionalen Klimamodelle CLM und REMO, diese deckten alle drei zu betrachtenden Perioden ab.

Das Pflanzenwachstumsmodell prognostizierte für alle Temperaturmodelle höhere Erträge in der ersten Periode (2021 - 2050) gegenüber den zwei späteren Perioden, was durch einen Anstieg der maximalen und minimalen Temperaturen sowie durch ein häufigeres Auftreten heißer Phasen während der Vegetationsperiode von der ersten hin zur letzten Temperaturperiode erklärt werden konnte. Die höchsten Erträge für die erste Temperaturperiode wurden mit dem STARII Temperaturmodell prognostiziert, weil dieses Modell das geringste Auftreten heißer Phasen im Verlauf der Vegetationsperiode annahm. Das CLM Modell hingegen ging in allen Temperaturperioden und an allen

Versuchsstandorten von einer deutlich höheren Anzahl an heißen Phasen mit deutlich höheren Temperaturen aus, wodurch das Pflanzenwachstumsmodell mit diesem regionalen Temperaturmodell CLM stets die niedrigsten Erträge prognostizierte.

Ziel dieser Arbeit war es, ein neues Pflanzenwachstumsmodell zu entwickeln, das die Entwicklung der Biomasse und des Ertrags von Winterweizen unter Verwendung von Temperatursummen und einfachen Partitionsregeln unter den in Bayern herrschenden Unterschieden in der Wasserversorgung und in den klimatischen Bedingungen simulieren sollte. Hierbei sollte ein möglichst geringer Dateninput durch den Benutzer vonnöten sein, um das Pflanzenwachstumsmodell einer möglichst großen Menge an Benutzern zugänglich zu machen. Aus diesem Grunde wurde das Pflanzenwachstumsmodell so konzeptioniert, dass es nicht nur in der Wissenschaft sondern auch in Bildungs- und in angewandten Bereichen benutzt werden kann.

7 References

Acock, B. and Allen, L.H.Jr., 1985. Crop responses to elevated carbon dioxide concentration. In: Direct Effects of Increasing Carbon Dioxide on Vegetation. DOE/ER-0238. B.R. Strain and J.D. Cure (eds.). US Department of Energy, Carbon Dioxide Research Division, Washington DC. pp. 53–97.

Acock, B.C., 1991. Modeling canopy photosynthetic response to carbon dioxide, light interception, temperature, and leaf traits. In Modeling Crop Photosynthesis-From Biochemistry to Canopy, ed. K. J. Boote & R. S. Loomis. CSSA special publication. 19, 41–56.

Addae, P.C., and Pearson, C.J., 1992. Thermal Requirements for Germination and Seedling Growth of Wheat. Australian Journal of Agricultural Research 43 (3), 585–594.

Alden, J. and Hermann, R.K., 1971. Aspects of the cold-hardiness mechanism in plants Botanical Review. 37 (1), 37–142.

Allen Jr.L.H., Kirkham, M.B., Olszyk, D.M., and Whitman C.E., (Eds.), 1997. Advances in Carbon Dioxide Research. ASA Special Publication 61, Madison, WI, 228 pp.

Allen, R.G., Pereira, L.S., Raes, D., and Smith, M., 1998. Crop evapotranspiration - Guidelines for computing crop water requirements - FAO Irrigation and drainage. FAO - Food and Agriculture Organization of the United Nations. Rome, 56.

Allen, R.G., Pereira, L.S., Raes, D., and Smith, M., 2000. Crop evapotranspiration - Guidelines for computing crop water requirements - FAO Irrigation and drainage. FAO - Food and Agriculture Organization of the United Nations. Rome, 56. Reprinted 2000.

Alvarado V, and Bradford KJ., 2002. A hydrothermal time model explains the cardinal temperatures for seed germination. Plant, Cell and Environment. 25 (8), 1061–1069.

Andrews, J., Pomeroy, M.K., Seaman, W.L., Butler, G., Bonn, P.C., and Hoekstra, G., 1997. Relationships between planting date, winter survival and stress tolerances of soft white winter wheat in eastern Ontario. Canadian Journal of Plant Science. 77 (4), 507–513.

Andrews, J.E., 1960. Cold hardiness of sprouting wheat as affected by duration of hardening and hardening temperature. Canadian Journal of Plant Science. 40 (1), 93–102.

Angus, J.F., MacKenzie, D.H., Morton, R., and Schafer, C.A., 1981. Phasic Development in Field Crops. II. Thermal and Photoperiodic Responses of Spring Wheat. Field Crops Research. 4 (3), 269–283.

ArcGIS© 2010. ESRI® ArcMap™ 10.0, License Type: ArcInfo, Copyright© 1999-2010 ESRI Inc.

Ashworth, E.N., and Abeles, F.B., 1984. Freezing behavior of water in small pores and the possible role in the freezing plant tissues. Plant Physiology. 76 (1), 201–204.

Baker, C.K., Pinter, P.J., Reginato, R.J., and Kanemasu, E.T., 1986. Effects of Temperatures on Leaf Appearance in Spring and Winter Wheat Cultivars. Agronomy Journal. 78 (4), 605–613.

Banath, C.L. and Single, W.V., 1976. Frost injury to wheat stems and grain production. Australian journal of agricultural research. 27 (6), 749–753.

Bardossy, A., 1997. Downscaling from GCM to local climate through stochastic linkages. Journal of Environmental Management. 49 (1), 7–17.

Barrow, E.M., and Semenov, M.A., 1995. Climate Change Scenarios with High Spatial and Temporal Resolution for Agricultural Applications. Oxford Journals, Life Science, Forestry. 68 (4), 349–360.

Batts, G.R., Morison, J.I.L., Ellis, R.H., Hadley, P., and Wheeler, T.R., 1997. Effects of CO_2 and Temperature on Growth and Yield of Crops of Winter Wheat over Four Seasons. European Journal of Agronomy. 7 (1-3), 43–52.

Bauer, A., and Black, A.L., 1990. Stubble Height Effect on Winter Wheat in the Northern Great Plains (USA), I. Soil Temperature, Cold Degree-hours, and Plant Population. Agronomy Journal. 82 (2), 195–199.

Bayerische Landesanstalt für Landwirtschaft (LfL). Versuchsergebnisse aus Bayern (2001-2009). Faktorieller Sortenversuch, WINTERWEIZEN. Bayerische Landesanstalt für Landwirtschaft. Institut für Pflanzenbau und Pflanzenzüchtung.

Bayerische Vermessungsverwaltung, 2011, Geobasisdaten, Digitales Höhenmodell (DGM 25), www.geodaten.bayern.de

Bayerisches Landesamt für Umwelt (LfU), www.lfu.bayern.de

Benech Arnold RL, Ghersa CM, Sanchez RA, Insausti P. 1990. Temperature effects on dormancy release and germination rate in Sorghum halepense (L.) Pers. seeds: a quantitative analysis. Weed Research. 30 (2), 81–89.

Biebl, R., 1962. Protoplasmatische okologie der Pflanzen. Wasser and Temperatur. Springer-Verlag, Wien. 344 pp.

Blum, A., and Sinmena, B., 1994. Wheat Seed Endosperm Utilization under Heat Stress and its Relation to Thermo tolerance in the Autotrophic Plant. Field Crops Research. 37 (3), 185-191.

Blumenthal, C.S., Barlow, E.W.R., and Wrigley, C.W., 1993. Growth Environment and Wheat Quality: The Effect of Heat Stress on Dough Properties and Gluten Proteins. Journal of Cereal Science. 18 (1), 3–21.

Blumenthal, C.S., Stone, P.J., Gras, P.W. Bekes, F., Clarke, B., Barlow, E.W.R., Appels, R., and Wrigley, C.W., 1998. Heat-Shock Protein 70 and Dough-Quality Changes Resulting from Heat Stress during Grain Filling in Wheat. Cereal Chemistry. 75 (1), 43–50.

Boden, T.A., Kaiser, D.P., Sepanski, R.J., Stoss, F.W. (Eds.), 1994. Atmospheric Carbon Dioxide. In: Trends '93: A Compendium of Data on Global Change, Band 2. Carbon Dioxide Information Analysis Center, World Data Center-A for Atmospheric Trace Gases, Environmental Sciences Division, Oak Ridge National Laboratory, Oak Ridge, TN, pp. 3–6.

Brader, L., 1986. Early agrometeorological crop yield assessment, Ch3: Description of the crop forecasting method based on agrometeorological information. FAO Plant production and protection paper 73. FAO - Food and Agriculture Organization of the United Nations. Rome, 73, 15–37.

Brooking I.R., 1996. The temperature response of vernalization in wheat – a developmental analysis. Annals of Botany 78, 507–512.

Brooking I.R., Jamieson P.D., 2002. Temperature and Photoperiod Response of Vernalization in Near-Isogenic Lines of Wheat. Field Crops Research. 79, 21–38.

Brooks, R.J., and Tobias, A.M., 1996. Choosing the Best Model: Level of Detail, Complexity and Model Performance´. Mathematical and Computer Modelling. 24 (4), 1–14.

Brooks, R.J., Semenov, M.A., and Jamieson, P.D., 2001. ´Simplifying Sirius: Sensitivity and Development of a Meta-model for Wheat Yield Prediction´. European Journal of Agronomy 14 (1), 43–60.

Bunting, A.H., Dennett, M.D., Elston, J., and Speed, C.B., 1982. Climate and Crop Distribution. In: Blaxter, K., and Fowden, L. (Eds.), Food, Nutrition and Climate. Applied Science Publishers, London, pp. 43–74.

Burkart, S., Manderscheid, R., and Weigel, H-J, 2004. Interactive effects of elevated atmospheric CO_2 concentrations and plant available soil water content on canopy evapotranspiration and conductance of spring wheat. European Journal of Agronomy. 21 (4), 401–417.

Cao, W., and Moss, D.N., 1989. Temperature Effect on Leaf Emergence and Phyllochron in Wheat and Barley. Crop Science. 29 (4), 1018–1021.

Carsel, R.F., and Parrish, R.S., 1988. Developing joint probability distributions of soil water retention characteristics. Water Resources Research. 24 (5), 755–769.

Challinor, A.J., Wheeler, T.R., Craufurd, P.Q., Slingo, J.M., Grimes, D.I.F., 2004. Design and Optimization of a Large-area Process-based Model for Annual Crops. Agricultural and Forest Meteorology. 124 (1–2), 99–120.

Charles-Edwards, D.A., Doley, D., Rimmington, G.M., 1986. Modeling Plant Growth and Development. Academic Press, Sydney, 234 p.

Chaudhuri, U.N., Kirkam, M.B., Kanemasu, E.T., 1990. Root growth of winter wheat under elevated carbon dioxide and drought. Crop Science. 30 (4), 853–857.

Chen, T.H., Gusta, L.V., Fowler, D.B., 1983. Freezing injury and root development in winter cereals. Plant Physiology. 73 (3), 773–777.

Chouard. P., 1960. Vernalization and its relations to dormancy. Annual Review of Plant Physiology. 11, 191–238.

Conroy, J.P., Seneweera, S., Basra, A.S., Rogers, G., and Nissen Wooller, B., 1994. Influence of Rising Atmospheric CO_2 Concentrations and Temperature on Growth, Yield and Grain Quality of Cereal Crops. Australian Journal of Plant Physiology. 21 (6), 741–758.

Corbellini, M, Canevar, M.G., Mazza, L, Ciaffi, M., Lafiandra, D., and Borghi, B., 1997. Effect of the Duration and Intensity of Heat Shock during Grain Filling on Dry Matter and Protein Accumulation, Technological Quality and Composition in Bread and Durum Wheat. Australian Journal of Plant Physiology. 24 (2), 245–260.

Cornelis, W.M., Ronsyn, J., Van Meirvenne, M., and Hartmann, R., 2001. Evaluation of Pedotransfer functions for predicting the soil moisture retention curve. Soil Science Society of America Journal. 65 (3), 638–648.

Covell S., Ellis R.H., Roberts E.H., and Summerfield R.J., 1986. The influence of temperature on seed germination rate in grain legumes. I. A comparison of chickpea, lentil, soyabean and cowpea at constant temperatures. Journal of Experimental Botany. 37 (5), 705–715.

Cromey, M.G., Wright, D.S.C., Boddington, H.J., 1998. Effects of frost during grain filling on wheat yield and grain structure. New Zealand Journal of Crop and Horticultural Science. 26 (4), 279–290.

Cure, J.D., Acock, B., 1986. Crop responses to carbon dioxide doubling: a literature survey. Agricultural and Forest Meteorology. 38 (1-3), 127–145.

Del Pozzo, A. H., Garcia-Huidobro, J., Novoa, R., and Villaseca, S., 1987. Relationship of base temperature to development of spring wheat. Experimental Agriculture 23 (1), 21–30.

Doorenbos, J. and Kassam, A.H. 1979. Yield response to water. FAO Irrigation and Drainage Paper No. 33, FAO, Rome, Italy. 193 pp.

Doorenbos, J. and Pruitt, W.O., 1977. Crop water requirements. Irrigation and Drainage Paper No. 24, (rev.) FAO, Rome, Italy. 144 p.

Drake, B., Gonz`alez-Meler, M., Long, S., 1997. More efficient plants: a consequence of rising atmospheric CO_2? Annual Review Plant Physiology and Plant Molecular Biology. 48 (1), 609–639.

Drozodov, S.N., Titov, A.F., Balagurova, N.I., and Kritenko, S.P., 1984. The Effect of Temperature on Cold and Heat Resistance of Growing Plants. II. Cold Resistance Species. Journal of Experimental Botany. 53 (160), 1603–1608.

Dubrovsky, M., Nemesova, I., and Kalvova, J., 2005. Uncertainties in Climate Change Scenarios for the Czech Republic. Climate Research. 29 (2), 139–156.

Ellis R.H., Butcher PD. 1988. The effects of priming and natural differences in quality amongst onion seed lots on the response of the rate of germination to temperature and the identification of the characteristics under genotypic control. Journal of Experimental Botany 39 (7), 935–950.

Ellis R.H., Covell S., Roberts E.H., Summerfield R.J., 1986. The influence of temperature on seed germination rate in grain legumes. II. Interspecific variation in chickpea (Cicer arietinum L.) at constant temperature. Journal of Experimental Botany. 37 (10), 1503–1515.

Ellsworth, D., 1999. CO_2 enrichment in a maturing pine forest: are CO_2 exchange and water status in the canopy affected? Plant, Cell and Environment. 22 (5), 461–472.

Entz, M.H., and Fowler, D.B., 1988. Critical Stress Periods Affecting Productivity of No-till Winter Wheat in Western Canada. Agronomy Journal. 80 (6), 987–992.

European Soil Data Center (ESDAC), 2012. European Soil Portal – Soil and Data Information Systems, Institute for Environmental and Sustainability, Land Resource Management Unit, Joint Research Center, European Commission. http://eusoils.jrc.ec.europa.eu/.

Evans, J.P. 2012. Regional Climate Modelling: The Future for Climate Change Impacts and Adaptation Research. Institute of Electrical and Electronics Engineers (IEEE), Climate consensus, Earth observations. http://www.ieee.org/index.html.

Evans, L.T., 1993. Crop Evolution, Adaptation and Yield. Cambridge University Press, Cambridge. 500 p.

Evans, L.T., Wardlaw, I.F. and Fischer, R.A. 1975. Wheat. In: L.T. Evans, (ed.), Crop Physiology. Cambridge University Press, Cambridge, UK. Pp. 101–149.

Ewert, F., 2004. Modelling plant responses to elevated CO_2: how important is leaf area index? Annals of Botany. 93 (6), 619–627.

Ewert, F., Rodriguez, D., Jamieson, P., Semenov, M.A., Mitchell, R.A.C., Goudriaan, J., Porter, J.R., Kimball, B.A., Pinter Jr., P.J., Manderscheid, R., Weigel, H.J., Fangmeier, A., Fereres, E., Villalobos, F., 2002. Effects of elevated CO_2 and drought on wheat: testing crop simulation models for different experimental and climatic conditions. Agriculture, Ecosystem and Environment. 93 (1-3), 249–266.

FAO. 1996. FAO Statistics Series No. 130. Food and Agriculture Organization of the United Nations, Rome, 49.

Flood. R.G. and Halloran, G.M., 1986. Genetics and physiology of vernalization response in wheat. Advances Agronomy. 39, 87–124.

Folwer, D.B., Limin, A.E., Wang, S.,-Y., and Ward, R.W., 1996. Relationship between low-temperature tolerance and vernalization response in wheat and rye. Canadian Journal of Plant Science, 76 (1), 37–42.

Garcia-Huidobro J., Monteith J.L., Squire G.R., 1982a. Time, temperature and germination of pearl millet (Pennisetum typhoides S & H.). I. Constant temperature. Journal of Experimental Botany. 33 (2), 288–296.

Garcia-Huidobro J., Monteith J.L., Squire GR., 1982b. Time, temperature and germination of pearl millet (Pennisetum typhoides S & H.). II. Alternating temperature. Journal of Experimental Botany, 33 (2), 297–302.

Gerwitz, A. and Page, E.R., 1974. An empirical mathematical model to describe plant root systems. Journal of Applied Ecology. 11 (2), 773-781.

Goldstein, J., Gachon, P., Milton, J. and Parishkura, D., 2004. Statistical downscaling models evaluation: A regional case study for Quebec regions, Canada. Geophysical Research Abstracts.

Goudriaan, J., and van Laar, H.H. 1994. Modelling Potential Crop Growth Processes. Ch3: Climatic factors, Ch4: Assimilate flow and respiration. Department of Theoretical Production Ecology, Wageningen Agricultural University, Wageningen, The Netherlands. Kluwer Academic Publishers. The Netherlands.

Grace, J., 1988. Temperature as a Determinant of Plant Productivity. In: Long Sp, Woodward FI, (Eds.). Plants and Temperature. Cambridge: Society of Experimental Biology, Company of Biologists. 42, 91–107.

Gusta, L.V., Fowler, D.B. and Tyler, N.J.1982. Factors influencing hardening and survival in winter wheat. Pages 23–40 in P. H. Li and A. Sakai, eds. Plant cold hardiness and freezing stress. Vol. II. Academic Press, New York, NY.

Halevy, A.H., 1985, CRC Handbook of Flowering, vol. IV. CRC Press, Boca Raton.

Ham, J., Owensby, C., Coyne, P., Bremer, D., 1995. Fluxes of CO_2 and water vapor from a prairie ecosystem exposed to ambient and elevated atmospheric CO_2. Agricultural and Forest Meteorology. 77 (1-2), 73–93.

Hansen, J., Fung, I., Lacis, A., Rind, D., Lebedeff, S., Ruedy, R., Russell G. and Stone, P., 1988. Global climate changes as forecast by Goddard Institute for Space Studies three-dimensional model. Journal of Geophysical Research. 93 (D8), 9341–9364.

Hansen, J.W., 2002. Realizing the Potential Benefits of Climate Prediction of Agriculture: Issues, Approaches, Challenges. Agricultural Systems. 74 (3), 309–330.

Hanslin, H.M., and Mortensen, L.M., 2010. Autumn growth and cold hardening of winter wheat under simulated climate change. Acta Agriculturae Scandinavica section B–Soil and Plant Science. 60 (5), 437–449.

Hardegree S.P., Van Vactor S.S., Pierson F.B., and Palmquist D.E., 1999. Predicting variable-temperature response of non-dormant seeds from constant-temperature germination data. Journal of Range Management. 52, 83–91.

Hardegree, S. P., 2006. 'Predicting Germination Response to Temperature. I. Cardinal-temperature Model and Subpopulation-specific Regression'. Annals of Botany. 97 (6), 1115–1125.

Hileman, D., Huluka, G., Kenjige, P., Sinha, N., Bhattacharya, N., Biswas, P., Lewin, K., Nagy, J. and Hendrey, G., 1994. Canopy photosynthesis and transpiration of field-grown cotton exposed to free-air CO_2 enrichment (FACE) and differential irrigation. Agricultural and Forest Meteorology. 70 (1-4), 189–207.

Houghton, J.T., Jenkins, G.J. and Ephraums, J.J., 1990. Climate Change: The IPCC Scientific Assessment. Cambridge University Press, Cambridge.

Houghton, J.T., Meira Filho, L.G., Callander, B.A., Harris, N., Kattenberg, A., and Maskell, K., 1996. Climate Change 1995: The Science of Climate Change. Cambridge University Press, Cambridge, 572 p.

Hui, D., Luo, Y., Cheng, W., Coleman, J., Johnson, D., Sims, D., 2001. Canopy radiation- and water-use efficiencies as affected by elevated [CO_2]. Global Change Biology. 7 (1), 75–91.

Hunsaker, D.J, Kimball, B.A., Pinter Jr. P.J., Wall, G.W., LaMorte, R.L., Adamsen, F.J., Lwavitt, S.W., Thompson, T.L., Matthias, A.D., and Brooks, T.J, 2000. CO_2 enrichment and soil nitrogen effects on wheat evapotranspiration and water use efficiency. Agricultural and Forest Meteorology. 104 (2), 85–105.

Hunt, L. A., van der Poorten, D., and Pararajasingham, S., 1991. 'Post-anthesis Temperature Effects on Duration and Rate of Grain-filling in Some Winter and Spring Wheats'. Canadian Journal of Plant Science. 71 (3), 609–617.

Idso, S.B., 1989. Carbon dioxide and global change: Earth in transition. Arizona: IBR Press.

Iglesias, A., 2006. Use of DSSAT models for climate change impact assessment: Calibration and validation of CERES-Wheat and CERES-Maize in Spain. Universidad Politecnica de Madrid, Annex 1, 2 and 3. Contribution to: CGE Hands-on Training Workshop on V&A Assessment of the Asia and the Pacific Location Jakarta. Spain.

Intergovernmental Panel on Climate Change (IPCC), 1990. Scientific assessment of climate change. The Policymakers' Summary of the Report of Working Group I to the Intergovernmental Panel on Climate Change. Houghton JT, ed. World Meteorological Organization.

Intergovernmental Panel on Climate Change (IPCC), 2001. IPCC Third Assessment Report. Climate Change, 12 May 2003. http://www.ipcc.ch/.

Intergovernmental Panel on Climate Change (IPCC), 2007. Climate Change 2007: The Physical Science Basis. Contribution of Working Group I to the Fourth Assessment Report of the Intergovernmental Panel on Climate Change [Solomon, S., D. Qin, M. Manning, Z. Chen, M. Marquis, K.B. Averyt, M.Tignor and H.L. Miller (eds.)]. Cambridge University Press, Cambridge, United Kingdom and New York, NY, USA.

Intergovernmental Panel on Climate Change (IPCC), 2011. IPCC Special Report on Renewable Energy Sources and Climate Change Mitigation. Prepared by Working Group

III of the Intergovernmental Panel on Climate Change [O. Edenhofer, R. Pichs-Madruga, Y. Sokona, K. Seyboth, P. Matschoss, S. Kadner, T. Zwickel, P. Eickemeier, G. Hansen, S. Schlömer, C. von Stechow (eds)]. Cambridge University Press, Cambridge, United Kingdom and New York, NY, USA, 1075 p.

Jamieson P.D., Brooking I.R., Porter J.R., Wilson D.R., 1995. Prediction of Leaf Appearance in Wheat: a Question of Temperature. Field Crops Research 41 (1), 35–44.

Jamieson, P.D., Brooking, I.R., Semenov, M.A., and Porter, J.R., 1998a. Making Sense of Wheat Development: A Critique of Methodology. Field Crops Research 55 (1-2), 117–127.

Jamieson, P.D., Brooking, I.R., Semenov, M.A., McMaster, G.S., White, J.W., and Porter, J.R., 2007. Reconciling Alternative Models of Phenological Development in Winter Wheat. Field Crops Research 103 (1), 36–41.

Jamieson, P.D., Porter, J.R., Goudriaan, J., Ritchie, J.T., Keulen, H., van Stol, W., 1998b. A comparison of the models AFRCWHEAT2, CERES-Wheat, Sirius, SU-CROS2 and SWHEAT with measurements from wheat grown under drought. Field Crops Research. 55 (1-2), 23–44.

Jamieson, P.D., Semenov, M.A., Brooking, I.R., Francis, G.S., 1998c. Sirius: a mechanistic model of wheat response to environmental variation. European Journal of Agronomy. 8 (3-4), 161–179.

Johnson, R.C., and Kanemasu, E.T., 1983. Yield and Development of Winter Wheat at Elevated Temperatures. Agronomy Journal. 75: 561–565.

Jones, P., Allen, L., Jones, J., Valle, R., 1985. Photosynthesis and transpiration responses of soybean canopies to short- and long-term CO_2 treatment. Agronomy Journal. 77: 119–126.

Jones, P.D., Murphy, J.M. and Noguer, M. 1995. Simulation of climate change over Europe using a nested regional–climate model, I: Assessment of control climate, including sensitivity to location of lateral boundaries. Quarterly Journal of the Royal Meteorological Society. 121 (526), 1413–1449.

Jones, R.J.A., Hiederer, R., Rusco, E., Loveland, P.J. and Montanarella, L. 2004. The map of organic carbon in topsoils in Europe, Version 1.2, September 2003: Explanation of Special Publication Ispra 2004 No.72 (S.P.I.04.72). European Soil Bureau Research Report No.17, EUR 21209 EN, 26pp. and 1 map in ISO B1 format. Office for Official Publications of the European Communities, Luxembourg.

Jones, R.J.A., Hiederer, R., Rusco, E., Loveland, P.J. and Montanarella, L. 2005. Estimating organic carbon in the soils of Europe for policy support. European Journal of Soil Science, October 2005, 56, p.655-671.

Kattenberg, A., Giorgi, F., Grassl, H., Meehl, G.A., Mitchell, J.F.B., Stouffer, R.J., Tokioka, T., Weaver, A.J. and Wigley, T.M.L., 1996. Climate models - projections of future climate, in Climate Change 1995: The Science of Climate Change, 285-357, (Eds Houghton, J.T., Filho, L.G.M., Callander, B.A., Harris, N., Kattenberg, A. and Maskell, K.) Cambridge University Press, Cambridge, UK.

Kätterer, T., Andrén, O., and jansson, P-E., 2006. Pedotransfer functions for estimating plant available water and bulk density in Swedish agricultural soils. Acta Agriculturae Scandinavica Section B-Soil and Plant Science. 56 (4), 263–276.

Katterer, T., Hansson, A., Andren, O., 1993. Wheat root biomass and nitrogen dynamics – effects of daily irrigation and fertilization. Plant Soil 151: 21–30.

Katz, R.W., Brown, B.G., 1992. Extreme Events in a Changing Climate – Variability is more Important than Averages. Climatic Change. 21 (3), 289–302.

Keeling, C.D., 1991. CO_2 emissions. Historical record. Global. In Trends 91: A Compendium of Data on Global Change, ed. T. A. Boden, R. J. Sepanski and F. W. Stoss. Carbon Dioxide Information Analysis Center, Oak Ridge National

Kersebaum, K.C., Nendel, C., Mirschel, W., Manderscheid, Weigel, H.-J., and Wenkel, K.-O., 2009. Testing different CO_2 response algorithms against a face crop rotation experiment and application for climate change impact assessment at different sites in Germany. Quarterly Journal of the Hungarian Meteorological Service. 113(1–2), 79–88.

Kimball, B., Lamorte, R., Pinter, P., Wall, G., Hunsaker, D., Adamsen, F., Leavitt, S., Thompson, T., Matthias, A., Brooks, T., 1999. Free-air CO_2 enrichment and soil nitrogen effects on energy balance and evapotranspiration of wheat. Water Resources Research. 35 (4), 1179–1190.

Kimball, B., Lamorte, R., Seay, R., Pinter, P., Rokey, R., Hunsaker, D., Dugas, W., Heuer, M., Mauney, J., Hendrey, G., Lewin, K., Nagy, J., 1994. Effects of free-air CO_2 enrichment on energy balance and evapotranspiration of cotton. Agricultural and Forest of Meteorolgy. 70 (1–7), 259–278.

Kimball, B.A., 1983. Carbon dioxide and agricultural yield: an assemblage and analysis of 430 prior observations. Agronomy Journal. 75, 779–86.

Kimball, B.A., Kobayashi, K., Bindi, M., 2002. Responses of agricultural crops to free-air CO_2 enrichment. Advances in Agronomy. 77, 293–368.

Kirby, E.J.M., 1985. Significant Stages of Ear Development in Winter Wheat. In: Day, W., and Aktin, R.K. (Eds.), Wheat Growth and Modeling. NATO ASI Series. Series A: Life Sciences, vol. 86, Plenum Press, London.

Kirby, E.J.M., 1993. A Template for the Analysis of Variety Differences in Cereals. Aspects of Applied Biology. 34, 49–78.

kirby, E.J.M., Evans, E.J., Frost, D.L., Sylvester-Bradley, R., Spink, J.H., Clare, R.W., Scott, R.K., and Foulkes, M.J., 1998. Volume VI of V. varietal responses to length of growing season. 85 pp. In: (Ed. M.J. Foulkes), HGCA project Report no. 166E. Exploitation of varieties for UK cereal production. Home-Grown cereals Authority, Caledonia House, 223 Pentonville Road, London NI 9NG.

Körner, C., 2006. Plant CO_2 responses: an issue of definition, time and resource supply. New Phytologist, 172 (3), 393–411.

Kruijt, B., Witte, J-P.M., Jacobs, C.M.J., and Kroon, T., 2008. Effects of rising atmospheric CO_2 on evapotranspiration and soil moisture: A practical approach for the Netherlands. Journal of Hydrology. 349 (3–4), 257–267.

Kümmel, B., 1997. Temp, Humidity & Dew Point ONA. Roskilde Universitätscenter. www.faqs.org/faqs/meteorology/temp-dewpoint/

Lawless, C., and Semenov, M.A., 2005. 'Assessing Lead-time for Prediction Wheat Growth Using a Crop Simulation Model'. Agricultural and Forest Meteorology 135 (1–4), 302–313.

Legates, D.R., and McCabe, G.J. 1999. Evaluating the use of "Goodness of Fit" Measures in Hydrologic and Hydroclimatic Model Validation. Water Resources Research 35 (1), 233–241.

Leij, F.j, Alves, W.J, van Genuchten, m.Th, and Williams, J.R., 1996. The UNSODA unsaturated soil hydraulic database, version 1.0, EPA Report EPA/600/R-96/095, EPA National Risk Management Laboratory, G-72, Cincinnati, OH, USA, http://www.ussl.ars.usda.gov/MODELS/UNSODA.htm.

Leonardos, E.D., Savitch, L.V., Huner, N.P.A., Oquist, G., and Grodzinski, B., 2003. Daily photosynthetic and C-export patterns in winter wheat laves during cold stress and acclimation. Physiologia Plantarum, 117 (4), 521–531.

Levitt, J., 1972. Responses of plants to environmental stresses. Academic Press, New York. 698 p. 1980. Responses of plants to environmental stresses. Volume I. Chilling, freezing and high temperature stress. Academic Press, New York-London, 2nd edition. 497 p.

Limin. A. E. and Fowler, D. B. 1985. Cold-hardiness response of sequential winter wheat segments to differing temperature regimes. Crop Science. 25 (5), 838–843.

Loss, S.P., 1987. Factors Affecting Frost Damage to Wheat in Western Australia. Division of Plant Research, Western Australian Department of Agriculture, South Perth. 30 p.

Lumsden, M.E., 1980. 'The Influence of Weather on Development of Winter Wheat'. Bsc report. University of Bath, UK.

MacDowell, F.D.H., 1973. Growth Kinetic of Marquis Wheat. 5. Morphogenic Dependence. Canadian Journal of Boanyt. 51 (7), 1259–1265.

MacLean, A.H., and Yager, T.U., 1972. Available water capacities of Zambian soils in relation to pressure plate measurements and particle size analysis. Soil Science. 113 (1), 23–29.

Mahfoozi, S., Limin, A.E., and Fowler, D.B., 2001a. Influence of vernalization and photoperiod responses on cold hardiness in winter cereals. Crop Science 41 (4), 1006–1011.

Mahfoozi, S., Limin, A.E., and Fowler, D.B., 2001b. Developmental regulation of low-temperature tolerance in winter wheat. Annals of Botany. 87 (6), 751–757.

Makan, J.R., Burke, J.J., and Orzach, K.A., 1987. The Thermal Kinetic Window as an Indicator of Optimum Plant Temperature. Plant and Cell Physiology Supplement. 83 (4), 87.

Marcellos, H., 1977. Wheat frost injury-freezing stress and photosynthesis. Australian Journal of Agricultural Research. 28 (4), 557–564.

Marcellos, H., and Single, W.V., 1972. The Influence of Cultivar, Temperature, and Photoperiod on Post-flowering Development of Wheat. Australian Journal of Agricultural Research. 23, 533–540.

McMaster G.S., Wilhelm, W.W., Palic D.B., Porter J.R., Jamieson P.D., 2003. Spring Wheat Leaf Appearance and Temperature: Extending the Paradigm?. Annals of Botany. 91 (6), 697–705.

McMaster, G.S., White, J.W, Hunt, L.A., Jamieson, P.D., 2008. Simulation the Influence of Vernalisation, Photoperiod and Optimum Temperature on Wheat Developmental Rates. Annals of Botany. 102 (4), 561–569.

McMaster, G.S., Wilhelm, W.W., and Morgan, J.A., 1992. Simulating Winter Wheat Shoot Apex Phenology. The Journal of Agricultural Science, Cambridge. 119 (1), 1–12.

Mearns, L.O., RosenZweig, C., Goldberg, R., 1997. Mean and Variance Change in Climate Scenarios: Methods Agriculture Application, and Measures of Uncertainty. Climate Change. 35, 367–396.

Microsoft SQL Server 2008, Microsoft SQL Server Management Studio, version 10.0.1600.22. © 2008 Microsoft corporation.

Microsoft Visual Studio 2010, version 10.0.30319.1 RTMRel, © 2010 Microsoft corporation.

Miglietta, F., 1989. Effect of Photoperiod and Temperature on Leaf Initiation Rates in Wheat (*Triticum spp.*). Field Crops Research. 21: 121–130.

Minansy, B., McBratney, A.B., 2000. Evaluation and development of hydraulic conductivity pedotransfer functions for Australian soil. Australian Journal of Soil Research. 38 (4), 905-926.

Mitchell, R., Mitchell, V., Lawlor, D., 2001. Response of wheat canopy CO_2 and water gas-exchange to soil water content under ambient and elevated CO_2. Global Change Biology. 7 (5), 599–611.

Moriasi, D.N., Arnold, J.G., Van Liew, M.W., Bingner, R.L., Harmel, R.D., Veith, T.L., 2007. Models evaluation Giudlines for Systematic Quantification of Accuracy in Watershed simulations, Transactions of the ASABE. 50 (3), 885–900.

Morison, J., 1998. Stomatal response to increased CO_2 concentration. Journal of Experimental Botany. 49, 443–452.

Morison, J.I.L., & Lawlor, D.W, 1999. Interactions between increasing CO_2 concentration and temperature on plant growth. Plant, Cell and Environment. 22, 659–682.

Morrison, M. J., McVetty, P. B. E., and Shaykewich, C. F. (1989). The determination and verification of a baseline temperature for the growth of Westar summer rape. Canadian Journal of Plant Science. 69 (2), 455–464.

Musich, V. N., Maistrenko, G. G., and Kolot, G. A., 1981. ′Dynamics of Frost Resistance in Winter Wheat in Relation to Temperature Regime′. Nauchno technicheskii Byulleten' Vsesoyuzonogo Selektsionno geneticheskogo Instituta. 1, 14–19.

Nash, J. E. and Sutcliffe, J. V., 1970. River flow forecasting through conceptual models part I – A discussion of principles, Journal of Hydrology. 10 (3), 282–290.

Nicholls, N., 1997. Increased Australian Wheat Yields due to Climate Trends. Nature 387 (6632), 484–485.

Nielsen, K.F., and Humphries, E.C., 1996. Effect of Temperature on Plant Growth. Soils Fertility. 29 (1), 1–7.

Olien. C.R. 1967. Freezing stresses and survival. Annual Review of Plant Physiology. 18, 387–408.

Pachepsky, Y.A., Timmlin, D. and Varallyay G., 1996. Artificial neural networks to estimate soil water retention from easily measurable data, Soil Science Society of America Journal. 60(3), 727–733.

Pachepsky, Ya.A., Shcherbakov, R.A., Várallyay, G., and Rajkai, K., 1982. Soil water retention as related to other soil physical properties. Pochvovedenie. 2, 42–52.

Pälchen, W. 1996. Soil Mapping. Issued by the Federal Institute for Geosciences and Natural Resources and the Geological Survey office in the Federal Republic of Germany. Ed: Finnern, H., Grottenthaler, W., Kühn, D., Pälchen, W., Schraps, W-G. and Sponagel, W. 4th improved and expanded edition. Hannover, Germany. In German.

Páldi, E., Rácz, I., and Lasztity, D., 1996. Effect of Low Temperature on the rRNA Processing in Wheat (*Triticum aestivum*). Journal of Plant Physiology. 148 (3–4), 374–377.

Panagos, P., Van Liedekerke, M., Jones, A., Montanarella, L. 2012. European Soil Data Centre: Response to European policy support and public data requirements. Land Use Policy, 29 (2): 329-338. doi:10.1016/j.landusepol.2011.07.003.

Panagos, P., Van Liedekerke, M., Montanarella, L., Jones, R.J.A. 2008. Soil organic carbon content indicators and web mapping applications, Environmental Modelling & Software, 23(9): 1207-1209.

Parry, M.L., 1990. Climate Change and World Agriculture. Earthscan, London.

Passioura, J.B., 1979. Accountability, philosophy and plant physiology. Search. 10, 347–350.

Paulsen, G.M. 1968. Effect of photoperiod and temperature on cold hardening in winter wheat. Crop Science. 8, 29–32.

Peiris, D.R., Crawford, J.W., Grashoff, C., Jefferies, R.A., Porter, J.R., and Marshall, B., 1996. A Simulation Study of Crop Growth and Development under Climate Change. Agricultural and Forest Meteorology. 79 (49), 271–287.

Penning de Vries, F.W.T., Laar, H.H. Van and Chardon, M.C.M., 1983. Bioenergetics of growth of seeds, fruits and storage organs. In: I. Yoshida (Ed.), Potential productivity of field crops under different environments. International Rice Research Institute, Los Baños, pp. 38–59.

Petr, J. 1991. Weather and Yield. Developments in Crop Science No. 20. Amsterdam, Elsevier. 288 p.

Pittman, H.A., 1933. Frost injury of wheat. Journal of Agricultural Western Australia. 10, 385–393.

Pomeroy, M.K., and Fowler, D.B., 1973. Use of Lethal Dose Temperature Estimates as Indices of Frost Tolerance for Wheat Cold Acclimated under Natural and Controlled Environments. Canadian Journal of Plant Science. 53 (3), 489–494.

Porter, J.R., 1993. AFRCWHEAT2: a model of the growth and development of wheat incorporating responses to water and nitrogen. European Journal of Agronomy. 2: 69–82.

Porter, J.R., and Gawith, M., 1999. Temperatures and the Growth and Development of Wheat: a Review'. European Journal of Agronomy. 10 (1), 23–36.

Porter, J.R., and Semenov, M.A., 1999. Climate Variability and Crop Yields in Europe. Nature 400 (6747), 724.

Porter, J.R., Kirby, E.J.M., Day, W., Adam, J.S., Appleyard, M., Ayling, S., Baker, C.K., Beale, P., Besford, R.K., Chapman, A., Fuller, M.P., Hampson, J., Hay. R.K.M., Hough, M., Matthews, S., Thompson, W.J., Weir, A.H., Willington, V.B.A., and Wood, D.W., 1987. An Analysis of Morphological Development Stages in Avalon Winter Wheat with Different Sowing Dates at 10 Sites in England and Scotland. The Journal of Agricultural Science. 109 (1), 107–121.

Porter, J.R., Semenov, M.A., 2005. Crop Responses to Climatic Variability. Philosophical Transactions of the Royal Society B-Biological science. 360 (1463), 2021–2035.

Poumadere, M., Mays, C., LeMer, S., Blong, R., 2005. The 2003 Heat Wave in France: Dangerous Climate Change Here and Now. Risk Analysis. 25 (6), 1483–1494.

Prášil, I.T., Prášilová, P., and Pánková, K., 2004. Relationships among vernalization, shoot apex development and frost tolerance in wheat. Annals of Botany. 94 (3), 413–418.

Priesack, E. 2005. EXPERT-N Version 3.0, Model Library Documentation. Institute of Soil Ecology. GSF. National Research Center. Copyright© Eckart Priesack 1996, 2002, 2005, 113 p.

Pritchard, S.G., and Rogers, H.H., 2000. Spatial and temporal deployment of crop roots in CO_2-enriched environments. New Phytologist, 147 (1), 55–71.

Probert R.J., 1992. The role of temperature in germination ecophysiology. In: Fenner M, ed. Seeds: the ecology of regeneration in plant communities. Wallingford, Oxon: CAB International, pp 285–325.

Prudhomme, C., Jakob, D. and Svensson, C., 2003. Uncertainty and climate change impact on the flood regime of small UK catchments. Journal of Hydrology. 277 (1–2), 1–23.

Racsko, P., Szeidl, L., and Semenov, M. A., 1991. A Serial Approach to Local Stochastic Weather Models. Ecological Modelling. 57 (1–2), 27–41.

Rajkai, K., and Várallyay, G., 1992. Estimating soil water retention from simpler properties by regression techniques. In: van Genuchten, M.Th., Leij, F., Lund, L.J. (Eds.). Methods for Estimating the Hydraulic Properties of Unsaturated Soils, Riverside, California, 11–13 October 1989, pp. 417–426.

Rawls, W.J., Ahuja, L.R., and Brakensiek, D.L., 1992. Estimating soil hydraulicproperties from soils data. In: van Genuchten, M.Th., et al. (Eds.). Indirect Methods for Estimating the Hydraulic Properties of Unsaturated Soils. Proceedings of International Workshop, Riverside, CA, October 11–13. 1989 University of California, Riverside, CA, pp. 329–340.

Rawls, W.J., and Brankiesiek, D.L., 1985. Prediction of soil water properties for hydrologic modeling. In: Jones, E.B., Ward, T.J. (Eds.). Watershed Management in Eighties. Proceedings of the symposium watershed management in the eighties, Denver, pp.293–299.

Rawson, H.M., and Macpherson, H.G. 2000. IRRIGATED WHEAT, Section 6: Explanations of plant development. Food and Agriculture Organization of the United Nations, Rome.

Reilly, J., Tubiello, F.N., McCarl, B., Melillo, J., 2001. Impacts of climate change and variability on agriculture. In: US National Assessment Foundation Document. National Assessment Synthesis Team, US Global Change Research Program, Washington, DC.

Renger, M., Bohne, K., Facklam, M., Harrach, T., Riek, W., Schäfer, W., Wessolek, G. and Zacharias, S. 2008. Results and proposals of the Working Group DBG "characteristics of the soil structure" for the estimation of soil physical characteristics. In German

Richardson, C.W., 1981. Stochastic Simulation of Daily Precipitation, Temperature, and Solar Radiation. Water Resources Research. 17 (1), 182–190.

Richardson, C.W., and Wright, D.A., 1984. WGEN: A Model for Generating Daily Weather Variables. US Department of Agriculture, Agricultural Research Service, ARS-8. USDA, Washington, DC. 38 p.

Ritchie, J.T., and Otter, S., 1985. Description and Performance of CERES-Wheat. A User-oriented Wheat Yield Model. ARS Wheat Yield Project, ARS. 38, 159–176.

Roberts, D.W.A., 1979. Duration of hardening and cold hardiness in winter wheat. Canadian Journal of Botany. 57 (14), 1511–1517.

Roberts, E.H., 1988. Temperature and Seed Germination. In: Long Sp, Woodward FI, (Eds.). Plants and Temperature. Cambridge: Society of Experimental Biology, Company of Biologists. 109–132.

Roberts, E.H., Qi, A., Ellis, R.H., Summerfield, R.J., Lawn, R.J., and Shanmugasundaram, S., 1996. Use of Field Observation to Characterise Genotypic Flowering Responses to Photoperiod and Temperature: A Soyabean Exemplar. Theoretical and Applied Genetics. 93 (4), 519–533.

Rosenzweig, C., Hillel, D., 1998. Climate Change and the Global Harvest. Oxford University Press, Oxford, UK.

Russell, G., and Wilson, G.W., 1994. An Agri-Pedo-Climatological-Knowledgr-Base of Wheat in Europe. Joint Research Center, European Commission, Luxembourg. 158 p.

Sanhewe, A.J., Ellis, R.H., Hong, T.D., Wheeler, T.R., Batts, G.R., Hadley, P., and Morison, J.I.L., 1996. The Effect of Temperature and CO_2 on Seed Quality Development in Wheat (*Triticum aestivum* L.). Journal of Experimental Botany. 47 (5), 631–637.

Schaap, M.G., 1999. Rosetta Version 1.0. US Salinity Laboratory, USDA, ARS: Riverside, CA, http://www.ussl.ars.usda.gov/rosetta/rosetta.htm.

Schaap, M.G., and Bouten, W., 1996. Modeling water retention curves of sandy soils using neural networks. Water Resources Research. 32 (10), 3033–3040.

Schaap, M.G., Leij, F.L. and van Genuchten, M.T., 1998. Neural network analysis for hierarchical prediction of soil hydraulic properties. Soil Science Society of America Journal. 62 (4), 847–855.

Semenov, M.A., 2007. Development of High Resolution UKCIP02-based Climate Change Scenarios in the UK. Agricultural and Forest Meteorology. 144 (1–2), 127–138.

Semenov, M.A., and Barrow, E.M., 1997. Use of Stochastic Weather Generator in the Development of Climate Change Scenarios. Climatic Change 35 (4), 397–414.

Semenov, M.A., and Brooks, R.J., 1999. Spatial Interpolation of the LARS-WG Stochastic Weather Generator in Great Britain. Climate Research 11 (2), 137-148.

Semenov, M.A., and Porter, J.R., 1995. Climatic Variability and the Modeling of Crop Yields. Agricultural and Forest Meteorology. 73 (3–4), 265–283.

Sexton, A.M., 2007. Evaluation of Swat Model Applicability for Waterbody Impairment Identification and TMDL Analysis. Dissertation of University of Maryland, Maryland, United States of America. 255 p.

Shein, E.V., Pachepsky, Ya.A., Guber, A.K., and Checkhova, T.I., 1995. Experimental determination of hydrophysical and hydrochemical parameters in mathematical models for moisture- and salt transfer in soils. Pochvovedenie. 11, 1479–1486.

Siegenthaler, U. and Sarmiento, J.L., 1993. Atmospheric carbon dioxide and the ocean. Nature, 365, 119–25.

Sinclair, T., and Seligman, N., 2000. Criteria for publishing papers on crop modeling. Field Crops Research. 68 (3), 165–172.

Singh, J., Knapp, H.V. and Demissie, M., 2004. Hydrologic Modeling of the Iroquois River Watershed using HSPF and SWAT [Online]. Available by Illinois State Water Survey, ISWS CR 2004-08 http://www.sws.uiuc.edu/pubdoc/CR/ISWSCR2004-08.pdf (verified July 25, 2007).

Single, W.V., 1964. Studies on frost injury to wheat. II. Ice formation within the plant. Australian Journal of Agricultural Research. 15 (5), 869–875.

Single, W.V., 1984. Frost injury and the physiology of the wheat plant. Farrer Memorial Orientation. The Journal of the Australian Institute of agricultural Science. 51 (2), 128–134.

Single, W.V., Marcellos, H., 1974. Studies on frost injury to wheat. IV. Freezing of ears after emergence from leaf sheath. Australian Journal of Agricultural Research. 25 (5), 679–686.

Skinner, H., 2007. Winter carbon dioxide fluxes in humid temperate pastures. Agricultural and Forest Meteorology, 144 (1–2), 32–43.

Slafer, G.A., and Rawason, H.M., 1994. Sensitivity of Wheat Phasic Development of Major Environmental Factors: a Re-examination of Some Assumptions Made by Physiologists and Modellers. Australian Journal of Plant Physiology. 21 (4), 393–426.

Slafer, G.A., and Rawason, H.M., 1995c. Base and Optimum Temperatures Vary with Genotype and Stage of Development in Wheat. Plant, Cell and Environment. 18 (6), 671–679.

Slafer, G.A., and Rawson, H.M., 1995a. Photoperiod × Temperature Interactions in Contrasting Wheat Genotypes: Time to Heading and Final Leaf Number. Field Crops Research. 44 (2–3), 73–83.

Slafer, G.A., and Rawson, H.M., 1995b. Rates and Cardinal Temperatures for Processes of Development in Wheat: Effect of Temperature and Thermal Amplitude. Australian Journal of Plant Physiology. 22 (6), 913–626.

Slafer, G.A., and Rawson, H.M., 1995d. Development of Wheat as Affected by Timing and Length of Exposure to Long Photoperiod. Journal or Experimental Botany. 46 (293), 1877–1886.

Slafer, G.A., and Savin, R., 1991. 'Developmental Base Temperature in Different Phenological Phases of Wheat (Triticum aestivum)'. Journal of Experimental Botany. 42 (241), 1077–1082.

SlideWrite Plus for windows, version 3.00. copyright© Advanced Graphics Software, Inc. 1985–1995.

Snyder, R.L., Lanini, B.J., Shaw, D.A., and Pruitt, W.O. 1989a. Using reference evapotranspiration (ET_o) and crop coefficients to estimate crop evapotranspiration (ETc) for agronomic crops, grasses, and vegetable crops. Cooperative Extension, Univ. California, Berkeley, CA, Leaflet No. 21427, 12 p.

Soil Survey Division Staff. 1993. Soil Survey Manual Handbook18, Soil Conservation Service, US Department of Agriculture, US Government Printing Office Washington, DC.

Spilde, L.A. 1989. Influence of seed size and test weight on several agronomic traits of barley and hard red spring wheat. Journal or Production Agriculture. 2 (2), 169–172.

Spink, J.H., kirby, E.J.M., Frost, D.L., Sylvester-Bradley, R., Scott, R.K., Foulkes, M.J., Clare, R.W., and Evans, E.J., 2000. Agronomic implications of variation in wheat development due to variety, sowing date, site and season. Plant Varieties and Seeds. 13 (2), 91–105.

Stapper, M., and Lilly, J.M., 2003. Evaluation of Simtag and Nwheat in Simulating Wheat Phenology in Southeastern Australia. Australia 10th Australian Agronomy Conference, Hobart . Australian Society of Agronomy.

Stone, P.J., and Nicolas, M.E., 1995. Effect of Timing of Heat Stress During of Grain-filling on Two Wheat Varieties Differing in Heat Tolerance. I. Grain Growth. Australian Journal of Plant Physiology. 22 (6), 927–934.

Tamari, S., Wösten, J.H.M., and Ruiz-Suarez, J.C., 1996. Testing an artificial neural network for predicting soil hydraulic conductivity. Soil Science Society of America Journal. 60 (6), 1732-1741.

Tester, R.F., Morrison, W.R., Ellis, R.H., Piggott, J.R., Batts, G.R., Wheeler, T.R., Morison, J.I.L., Hadley, P., and Ledward, D.A., 1995. Effects of Elevated Growth Temperature and Carbon Dioxide Levels on Some Physicochemical Properties of Wheat Starch. Journal of Cereal Science. 22 (1), 63–71.

Thornley, J.H.M., 1980. Research strategy in the plant sciences. Plant, Cell and Environment. 3: 233–236.

Trigo, R.M., Palutikof, J.P., 2001. Precipitation Scenarios over Iberia: A Comparison Between Direct GCM Output and Different Down-Scaling Techniques. Journal of Climate. 14 (23), 4422–4446.

Tubiello, F.N., and Ewert, F., 2002. Simulating the effects of elevated CO_2 on crops: approaches and applications for climate change. European Journal of Agronomy. 18 (1–2), 57–74.

Tubiello, F.N., Rosenzweig, C., Volk, T., 1995. Interactions of CO_2, temperature, and management practices: simulations with a modified CERES-Wheat model. Agricultural Systems. 49 (2), 135–152.

Van Genuchten, M.Th, 1980. A closed-form equation for predicting the hydraulic conductivity of unsaturated soils. Soil Science Society of America Journal. USA, 44 (5), 892–898.

van Laar, E.H.H., Goudriaan, J. and van Keulen, H., 1997. SUCREOS97: Simulation of crop growth for potential and water-limited production situations. As applied to spring wheat. Wageningen. DLO Research Institute for Agrobiology and Soil Fertility (AB-DLO); Wageningen: The C.T. de Wit Graduate School for Production Ecology (PE). Quantitative approaches in systems analysis, no. 14.

Vereecken, H., Maes, J., Feyen, J. and Darius, P., 1989. Estimating the soil moisture retention characteristics from texture, bulk density and carbon content. Soil Science. 148 (1989), 389–403.

Wallwork, M.A.B., Jenner, C.F., Lougue, S.J., and Sedgley, M., 1998. Effect of High Temperature during Grain-Filling on the Structure of Developing and Malted Barley Grains. Annals of Botany. 82 (5), 391–397.

Wang S.-Y., Ward, R.W., Ritchie, J.T., Fischer, R.A. and Schulthess, U., 1995. Vernalization in wheat I. A model based on the interchangeability of plant age and vernalization duration. Field Crop Research. 41 (2), 91–100.

Wardlaw, I.F. and Moncur, L., 1995. The Response of Wheat to High Temperature Following Anthesis, I. The Rate of Duration of Kernel Filling. Australian Journal of Plant Physiology. 22 (3), 391–397.

Wardlaw, I.F., 1974. Temperature Control of Translocation. Bulletin – Royal Society of New Zealand. 12, 533–538.

Warrick, B.E, and Miller, T., 1999. Freeze injury on wheat. Texas Agricultural Extension service. The Texas A&M University System.

WDCC, 2008. World data center for climate, The German Climate Computing Center (DKRZ), http://www.ngdc.noaa.gov/wdcmain.html, Hamburg.

Wechsung, G., Wechsung, F., Wall, G., Adamsen, F., Kimball, B., Pinter, P., Lamorte, R., Garcia, R., Kartschall, T., 1999. The effects of free-air CO_2 enrichment and soil water availability on spatial and seasonal patterns of wheat root growth. Global Change Biology. 5 (5), 519–529.

Weir, A.E., Bragg, P.L., Porter, J.R., and Rayner, J.H., 1984. A Winter Wheat Crop Simulation Model without Water and Nutrient Limitations. Journal of Agricultural Science. Cambridge. 102, 371–382.

Whaley, J.M., Kirby, E.J.M., Spink, J.H., Foulkes, M.J., and Sparkes, D.L., 2004. Frost damage to winter wheat in the UK: the effect of plant population density. European Journal of agronomy. UK. 21 (1), 105–115.

Wheeler, T.R., Batts, G.R., Ellis, R.H., Hadley, B., and Morison, J.I.L., 1996a. Growth and Yield of Winter Wheat (*Triticum aestivum* L.) Crops in Response to CO_2 and Temperature. Journal of Agricultural Science. Cambridge. 127, 37–48.

Wheeler, T.R., Craufurd, P.Q., Ellis, R.H., Porter, J.R., and Vara Prasad, P.V., 2000. Temperature Variability and the Yield of Annual Crops. Agriculture, Ecosystem and Environment. 82 (1–3), 159–167.

Wheeler, T.R., Hong, T.D., Ellis, R.H., Batts, G.R., Morison, J.I.L., and Hadley, P., 1996b. The Duration and Rate of Grain Growth, and Harvest Index, of Wheat (*Triticum aestivum* L.) Crops in Response to Temperature and CO_2. Journal of Experimental Botany. 47 (298), 623–630.

Whitmore, A.P. and Addiscott, T.M., 1987. A function for describing nitrogen uptake, dry matter and rooting by wheat crops. Plant and Soil. 101 (1), 51–60.

Wilby, R.L., Hassan, H. and Hanaki, K., 1998. Statistical downscaling of hydrometeorological variables using general circulation model output. Journal of Hydrology. 205 (1–2), 1–19.

Wilcox, B.P., Rawls, W.J. Brakensiek, D.L. and Wight, J.R. 1990. Predicting Runoff from Rangeland Catchments: A Comparison of Two Models. Water Resources Research. 26 (10), 2401–2410.

Wilks, D.S., 1992. Adapting Stochastic Weather Generation Algorithms for Climate Change Studies. Climate Change. 22 (1), 67–84.

Wilks, D.S., and Wibly R.L., 1999. The Weather Generation Game: A Review of Stochastic Weather Models. Progress in Physical Geography. 23 (3), 329–357.

Williams, J., Ross, P., and Bristow, K., 1992b. Prediction of Campbell water retention function from texture structure, and organic matter. In: van Genuchten, M.Th., Leij, F.J., and Lund, L.J. (Eds.). Indirect Methods for Estimating the Hydraulic Properties of Unsaturated Soils. Proceedings of the International Workshop on Indirect Methods for Estimating the Hydraulic Properties of Unsaturated Soils, Riverside, California, 11–13 October 1989, pp. 427–441.

Williams, R.D., Ahuja, L.R., and Nancy, J.W., 1992a. Comparison of methods to estimate soil water characteristics from limited texture, bulk density, and limited data. Soil Science. 153 (3), 172–184.

Wilsie, C.P., 1962. Crop Adaptation and Distribution. W.H. Freeman and Company, San Francisco 4, California. 448 p.

Wilson, K., Carlson, T., Bunce, J., 1999. Feedback significantly influences the simulated effect of CO_2 on seasonal evapotranspiration from two agricultural species. Global Change Biology. 5 (8), 903–917.

Wollenweber, B., Porter, J.R., and Schellberg, J., 2003. Lack of Interaction between High-Temperature Events at Vegetative and Reproductive Growth Stages in Wheat. Journal Agronomy and Crop Science. 189 (3), 142–150.

Wösten, J.H.M., Finke, P.A., and Jansen, M.J.W., 1995. Comparison of class and continuous pedotransfer functions to generate soil hydraulic characteristics. Geoderma. 66 (3–4), 227–237.

Wösten, J.H.M., Lilly, A., Nemes, A., Le Bas, C., 1999. Development and use of a database of hydraulic properties of European soils. Geoderma. 90 (3–4), 169–185.

8 Appendix

A1. The yield (———), leaves (———), stem (———), and straw (———) performance (Dry matter (g.dm/m^2.d)) at the studied seasons, 2000/01 (a), 2001/02 (b), 2002/03 (c), 2003/04 (d), 2004/05 (e), 2005/06 (f), 2006/07 (g), 2007/08 (h), and 2008/09 (i) at:

1. Donau-Ries

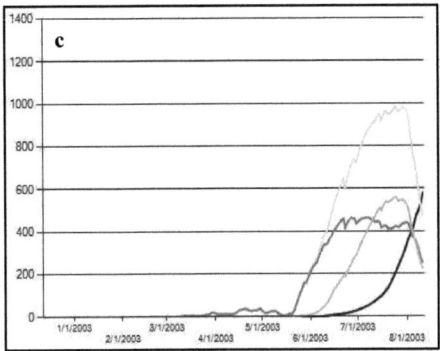
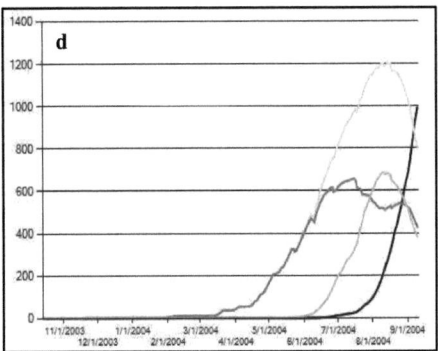
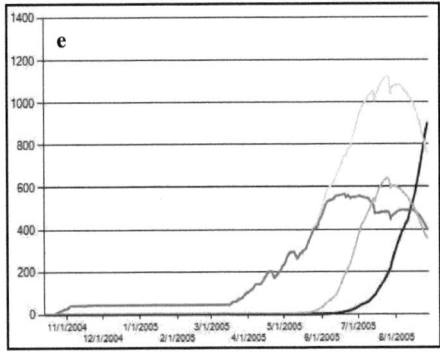
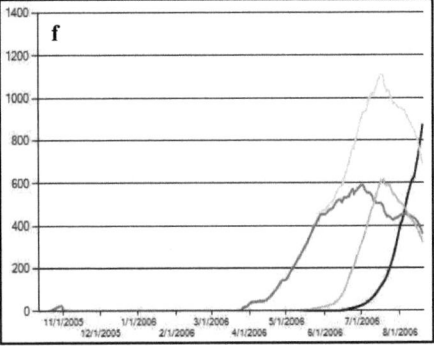

Date   Date

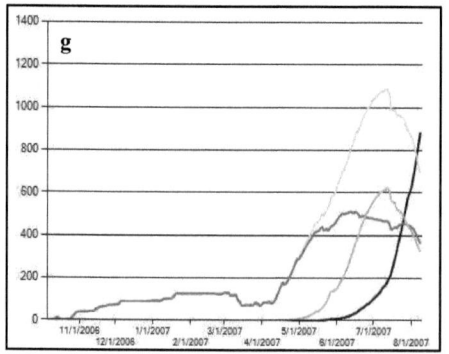

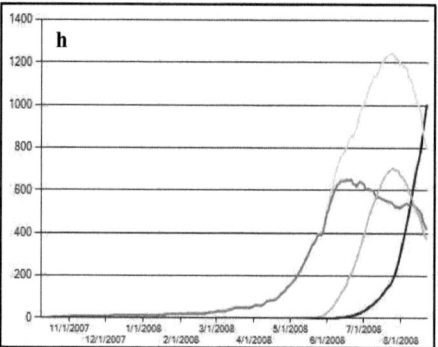

Date

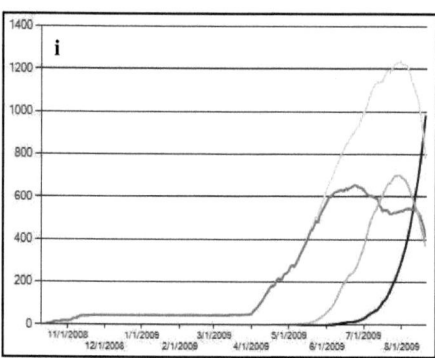

Date

2. Eichstätt

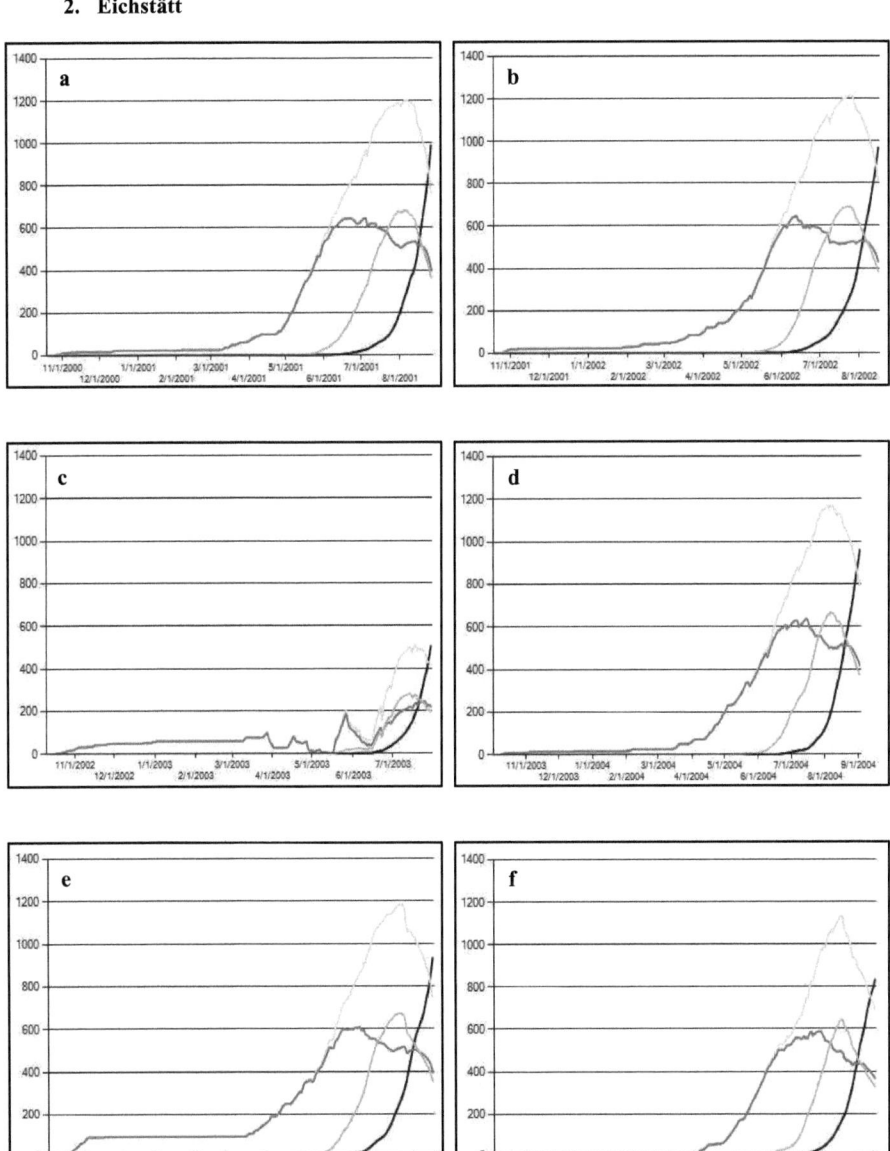

Date   Date

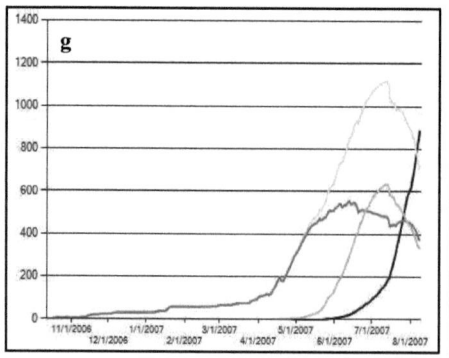 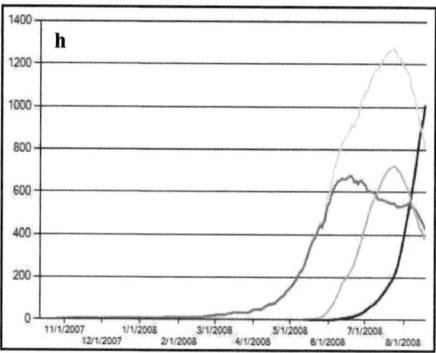

Date

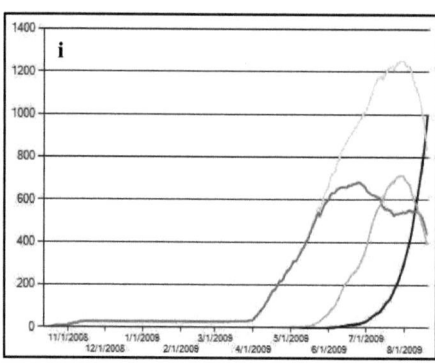

Date

3. Freising

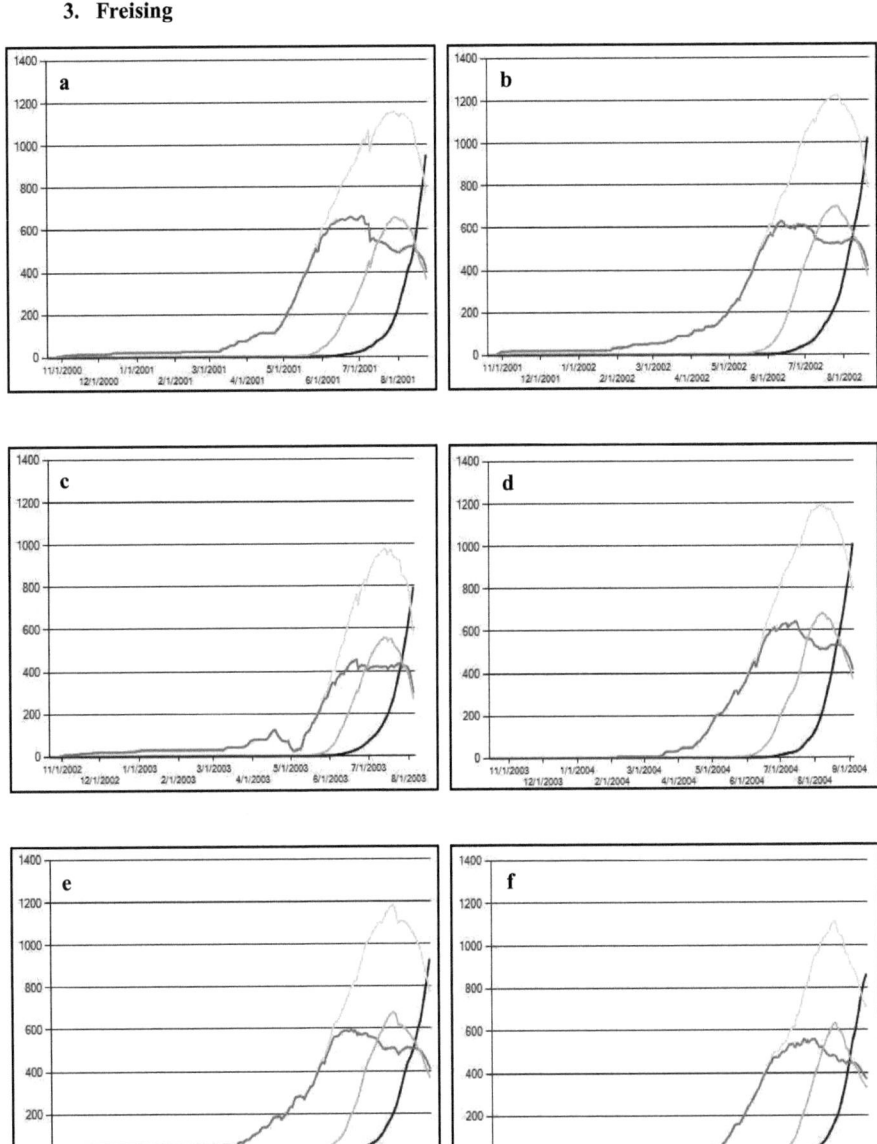

Date   Date

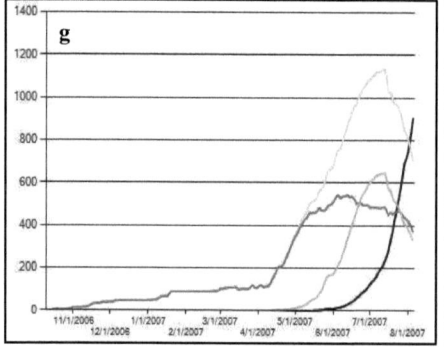
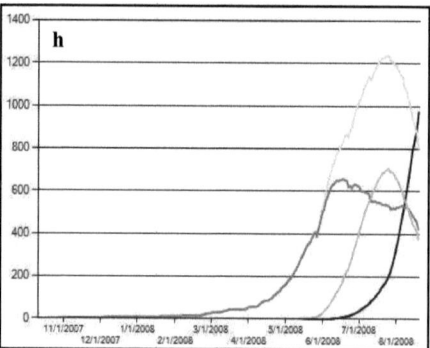

Date

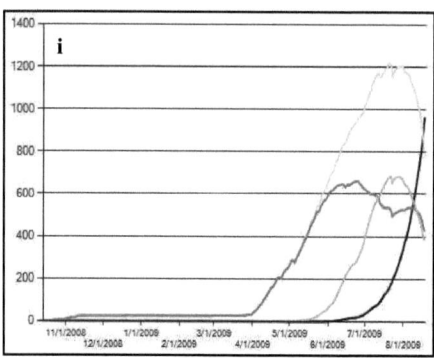

Date

4. **Günzburg**

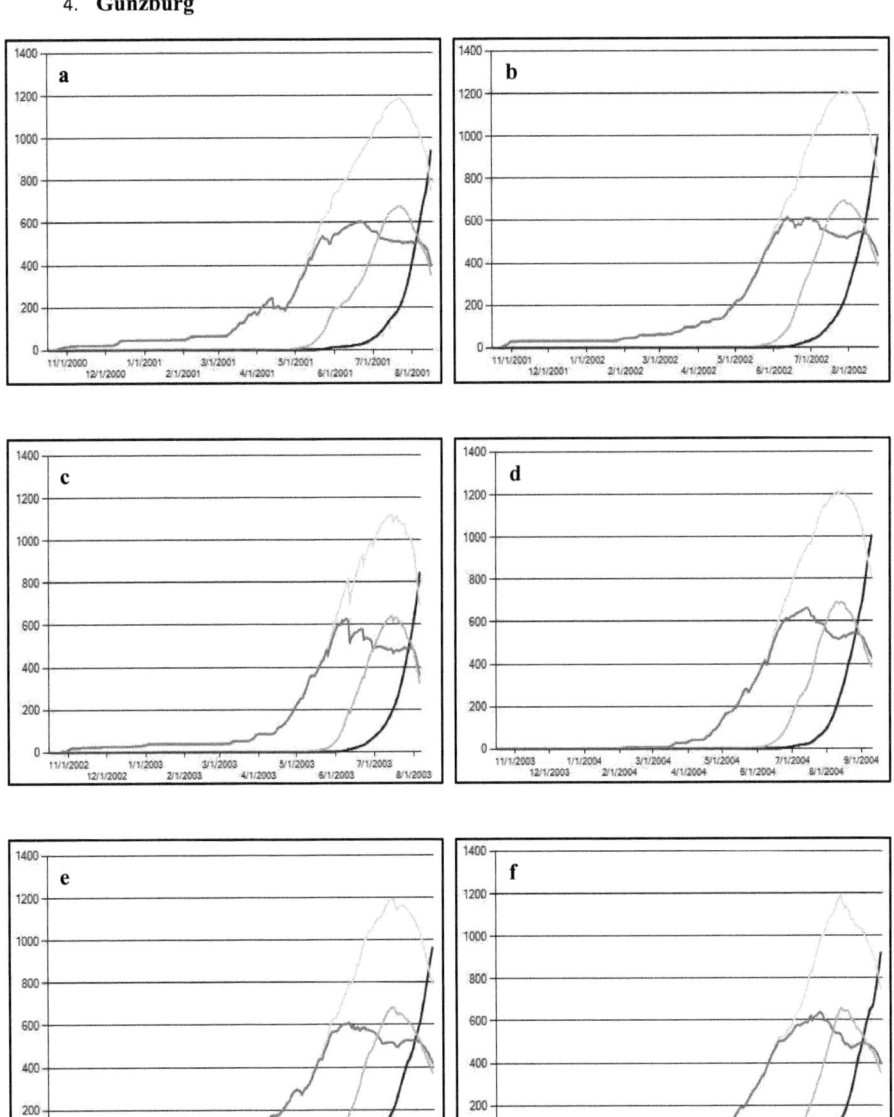

Date

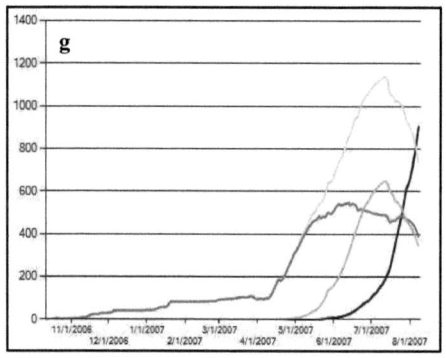
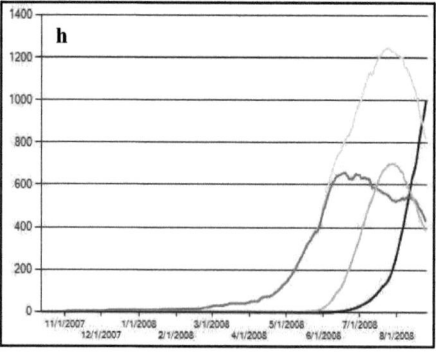

Date

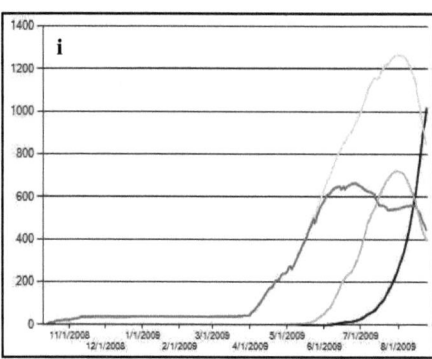

Date

5. Lichtenfels

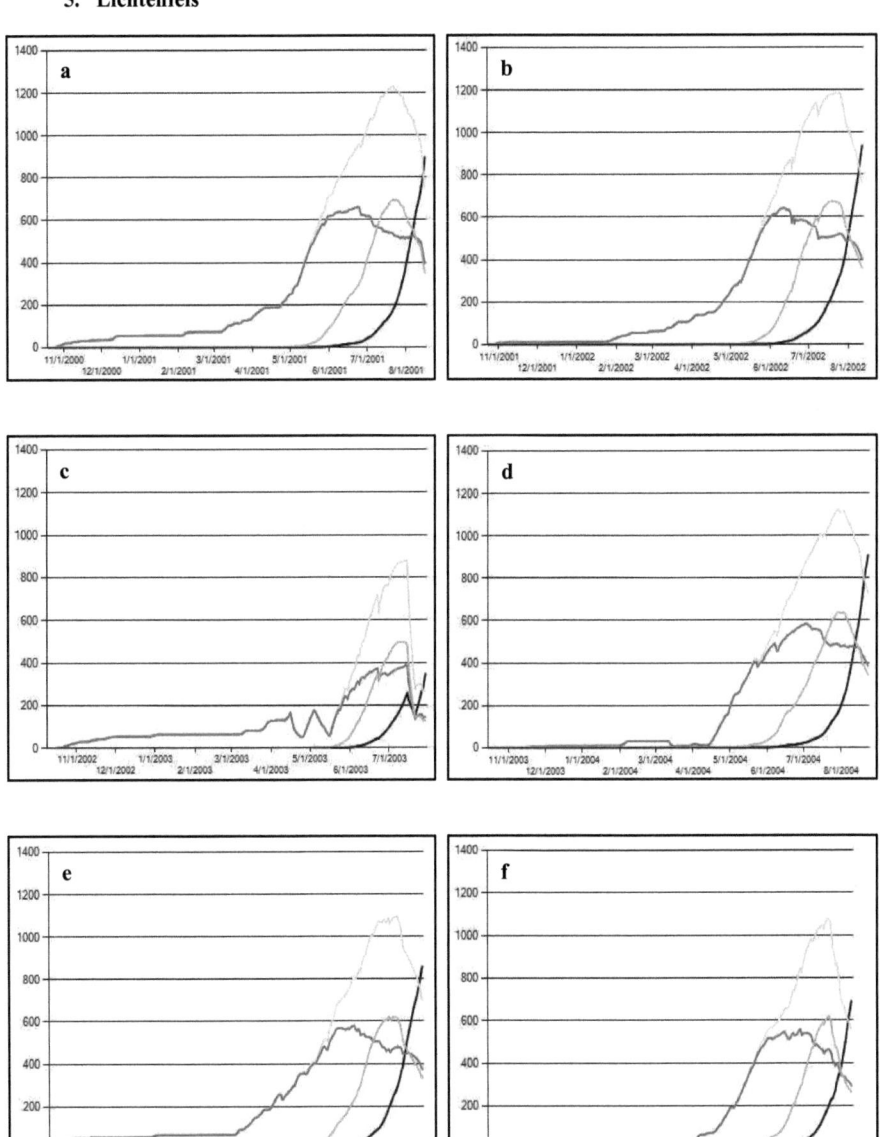

Date   Date

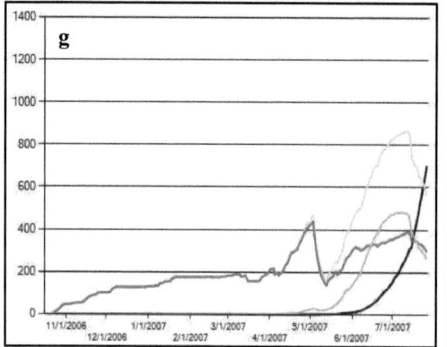
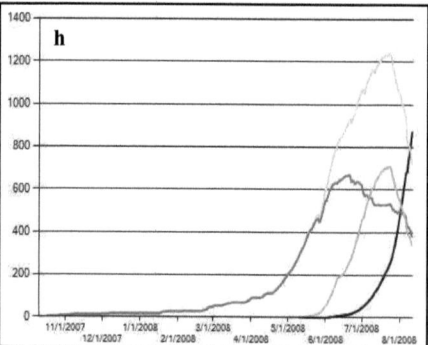

Date

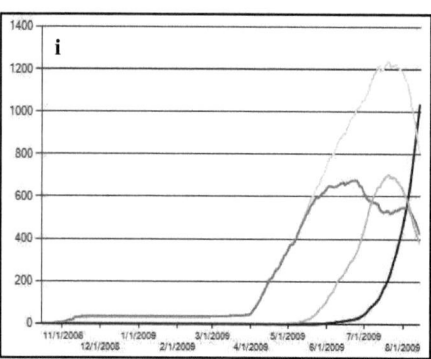

Date

6. Main-Spessart

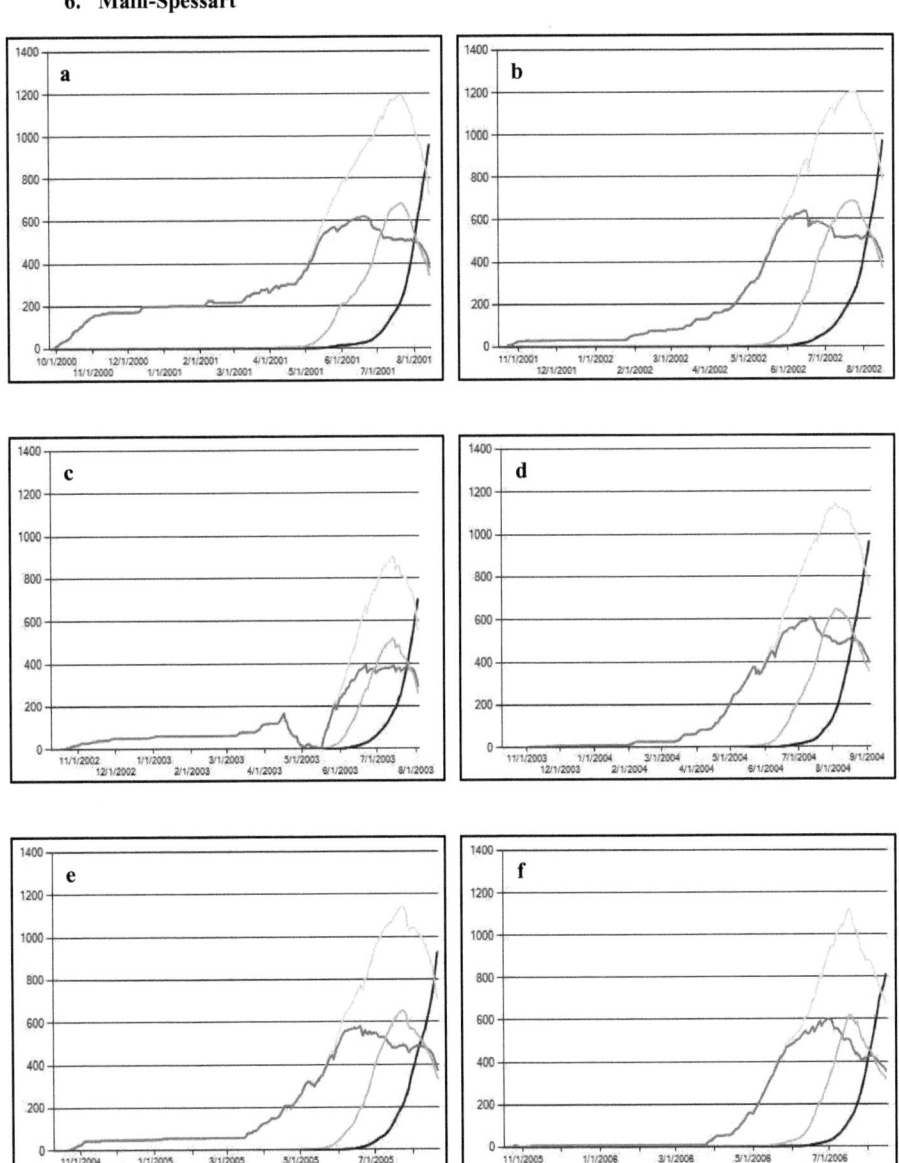

Date   Date

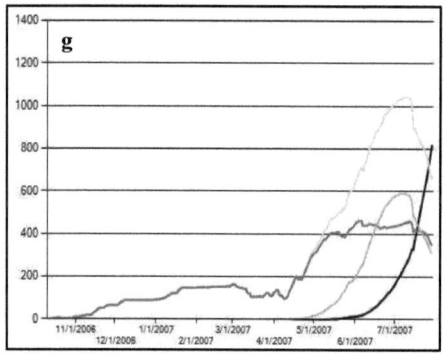
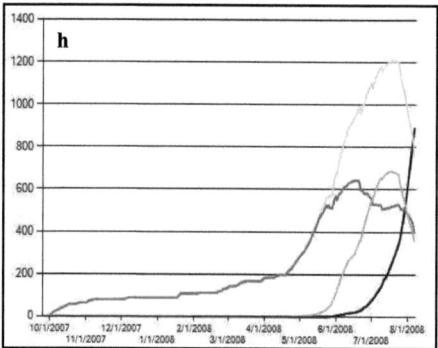

Date

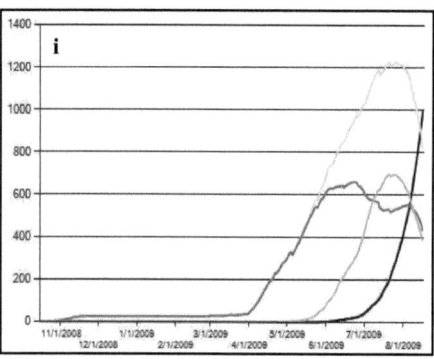

Date

7. Neumarkt

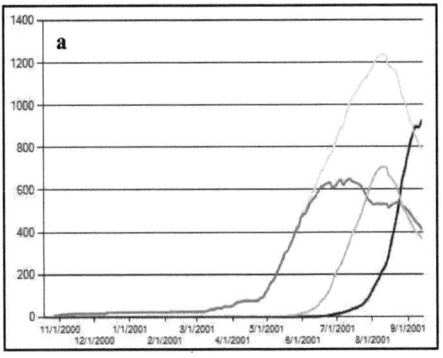
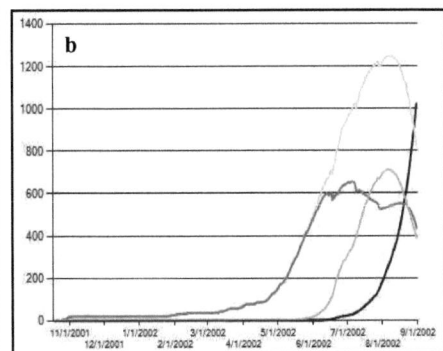
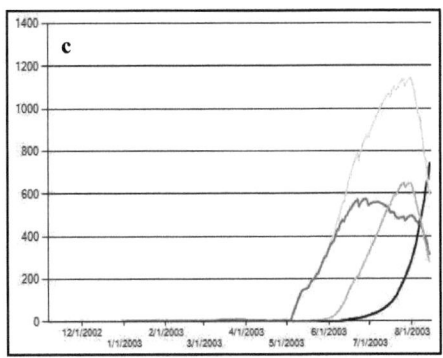
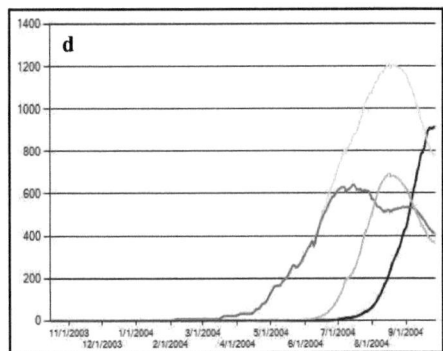
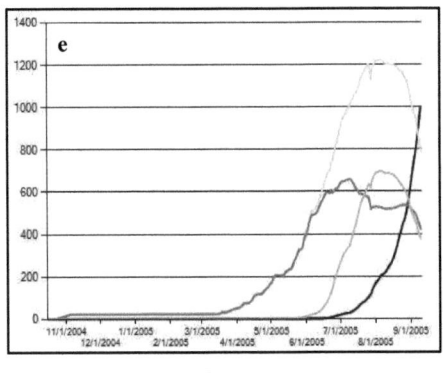
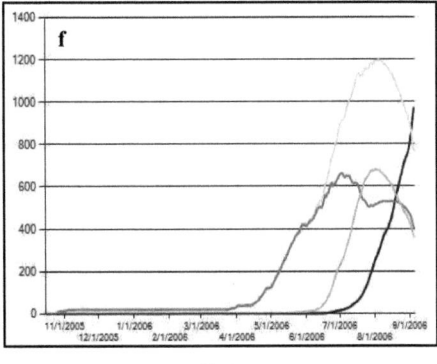

Date　　　　　　　　　　　　　　　Date

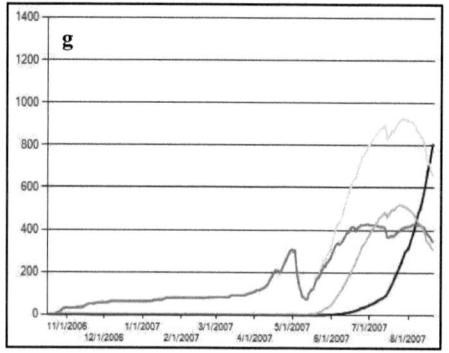

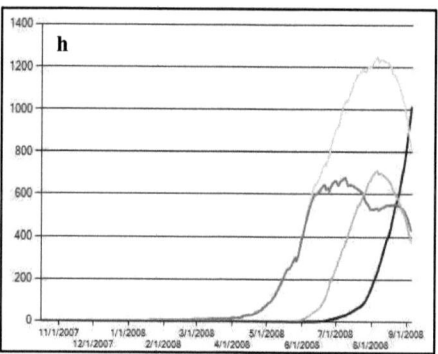

Date

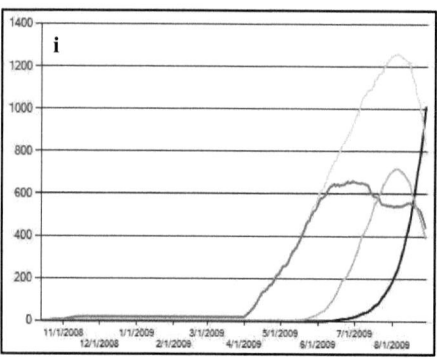

Date

8. Passau

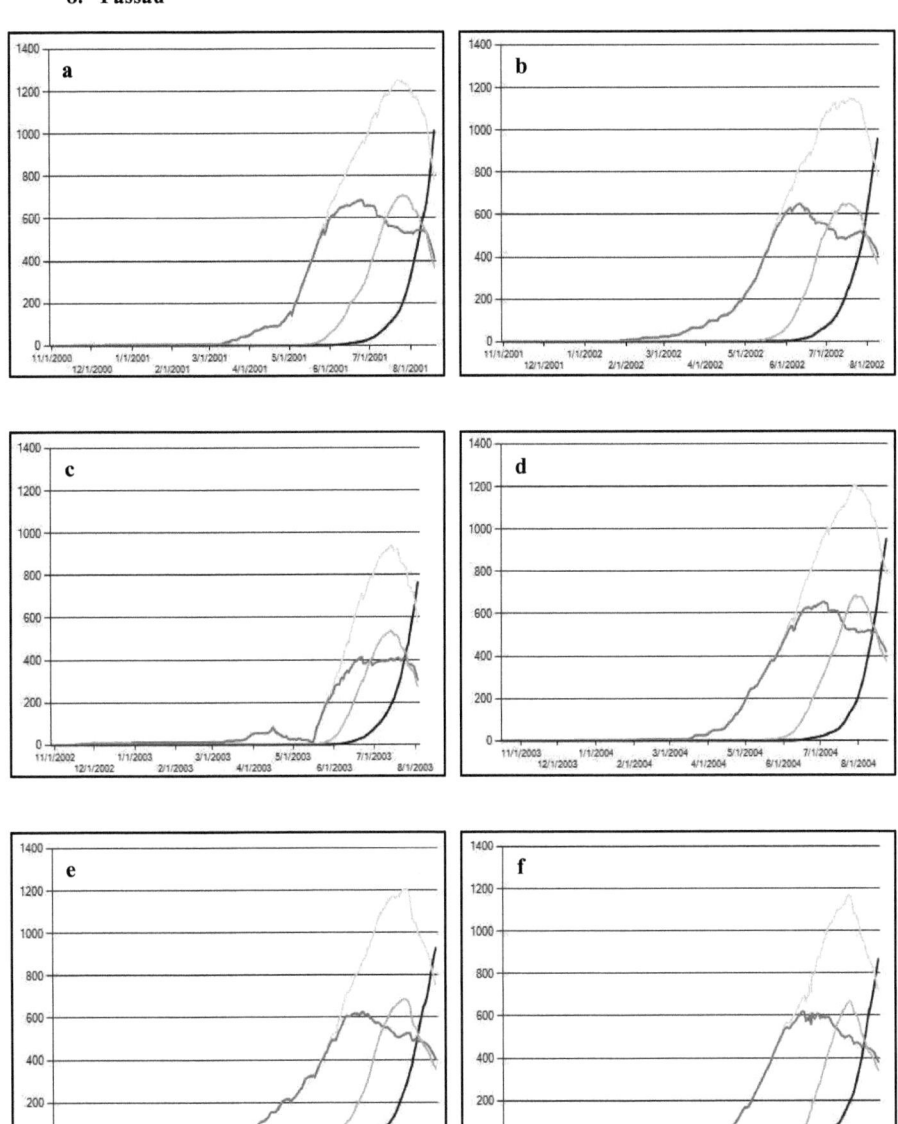

Date   Date

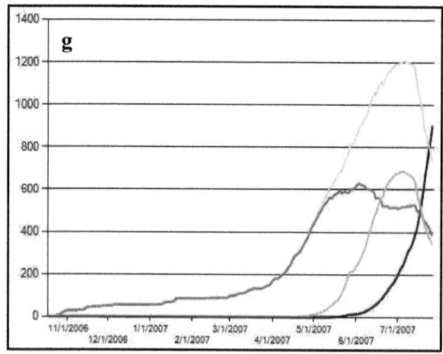
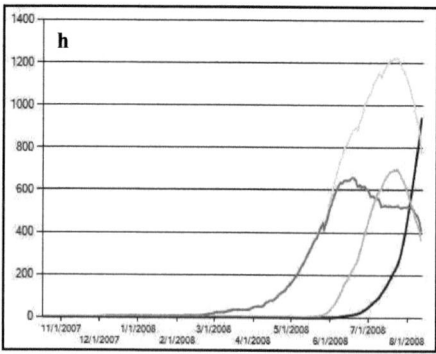

Date

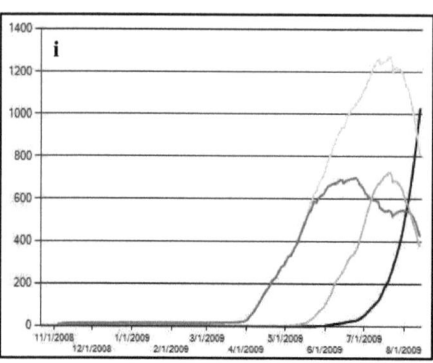

Date

9. Regensburg

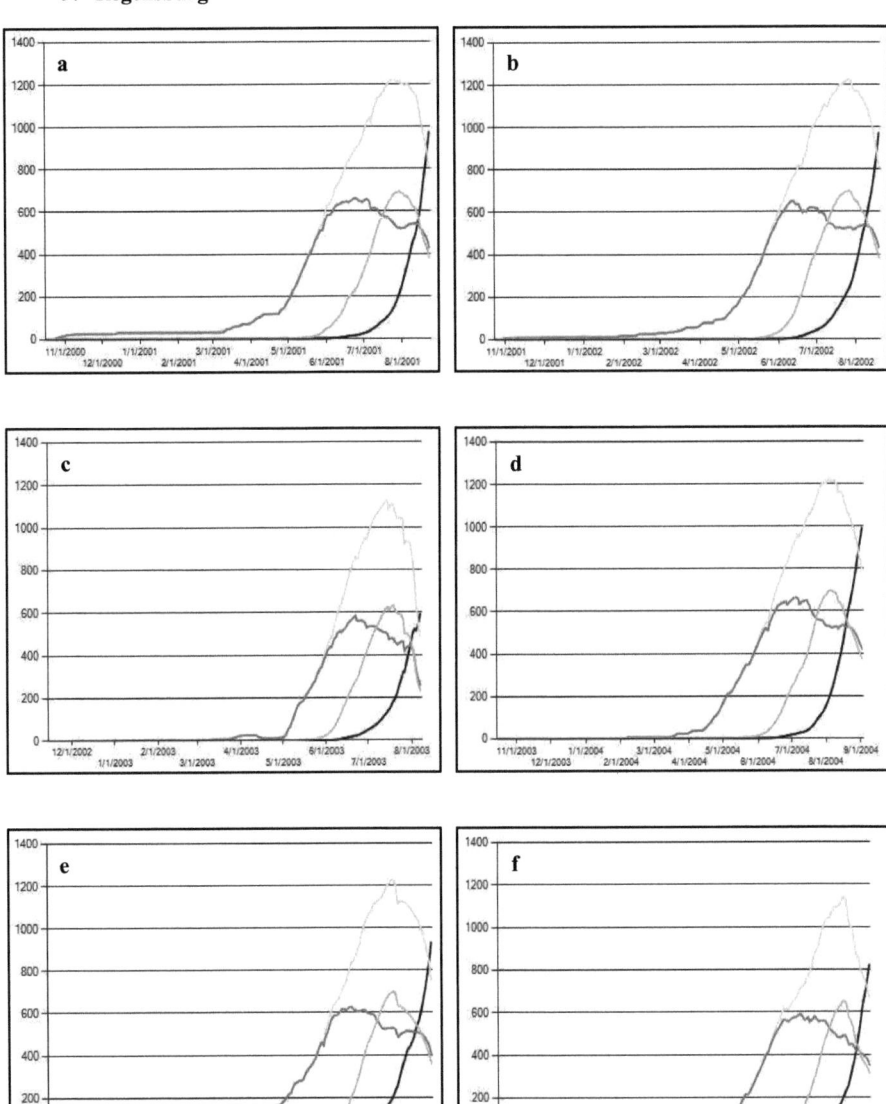

Date

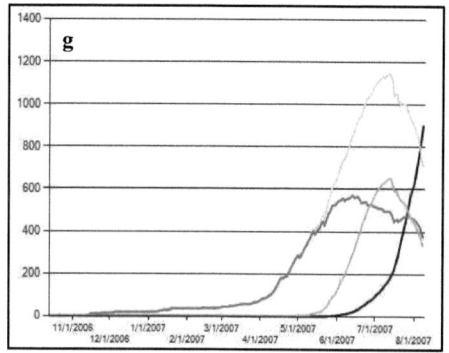
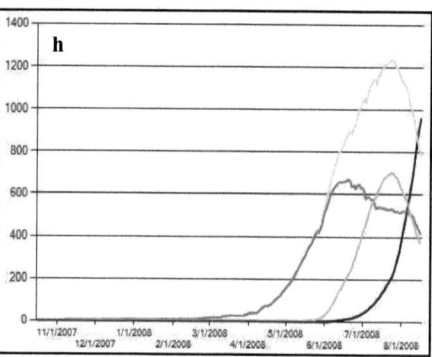

Date

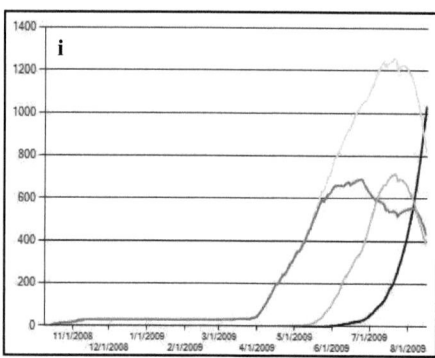

Date

10. Weißenburg-Gunzenhausen

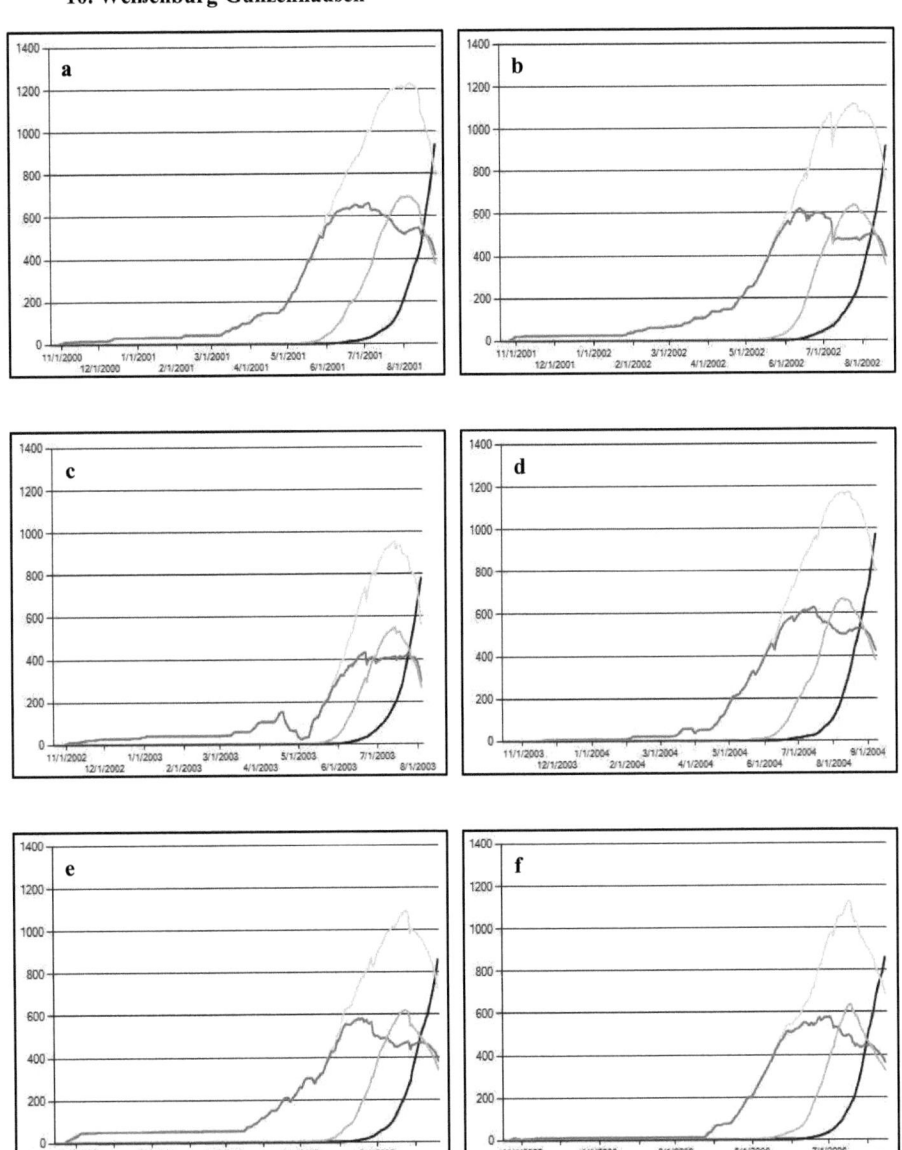

Date   Date

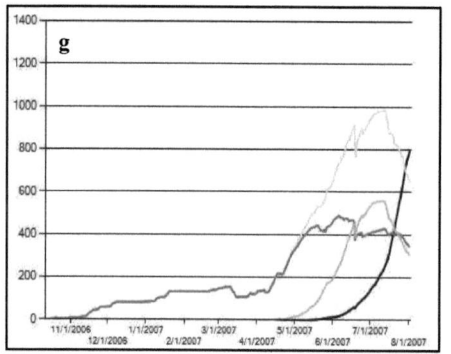
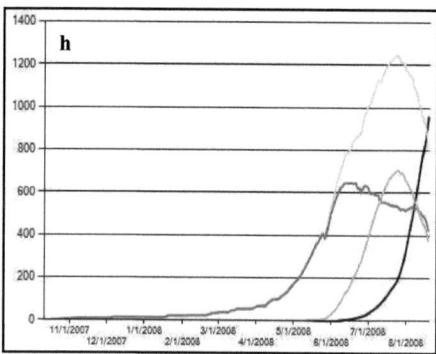

Date

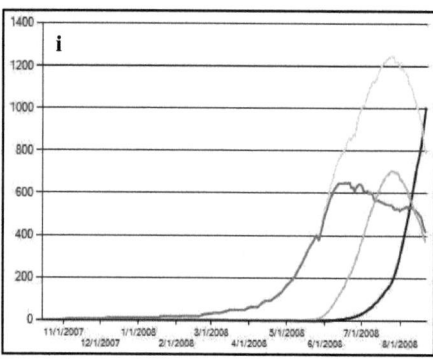

Date

11. Würzburg

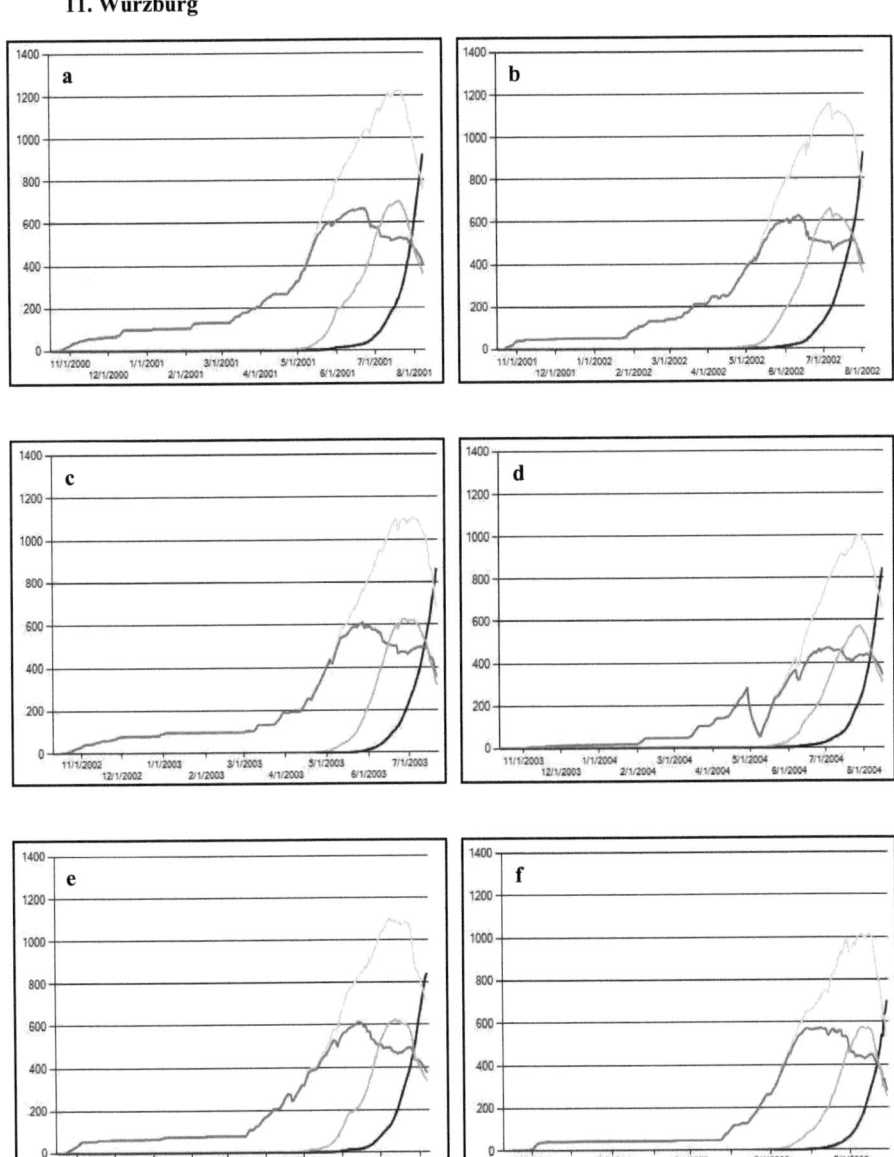

Date

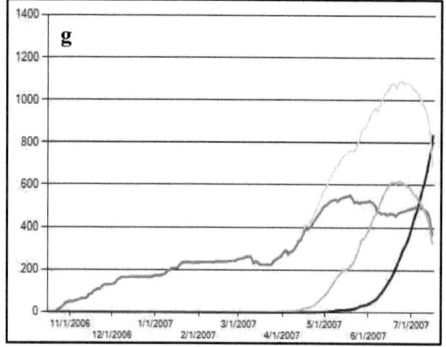

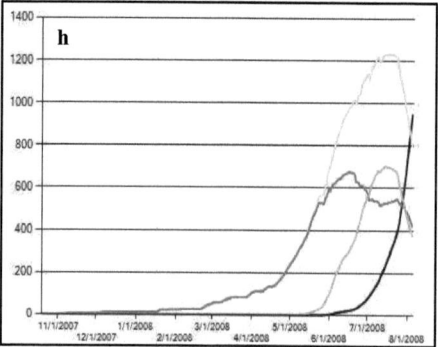

Date

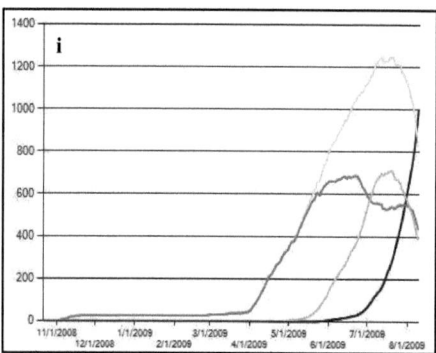

Date

A2. Comparison between the monthly average maximum temperature between the future predicted periods, 2021 – 2050 (— · ·), 2051 – 2080 (– – ·), and 2071 – 2100 (———) at at Würzburg (a), Donau-Ries (b), Freising (c), Passau (d), Regensburg (e), Lichtenfels (f), Eichstätt (g), Landshut (h), Neumarkt (i), Weissburg-Gunzenhausen (j), Main-Spessart (k), and Günzburg (l) for REMO model:

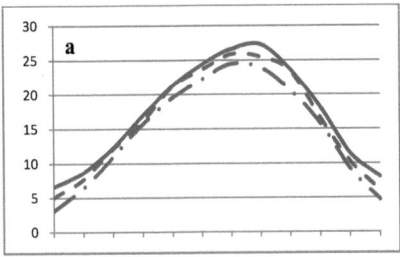
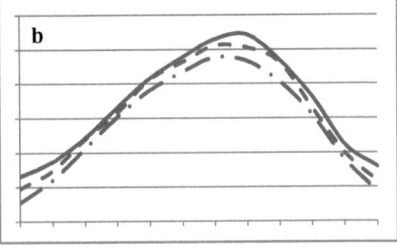
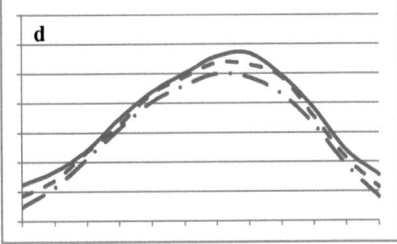
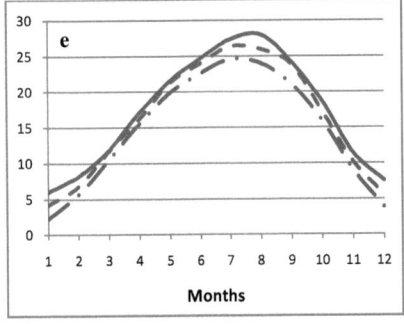
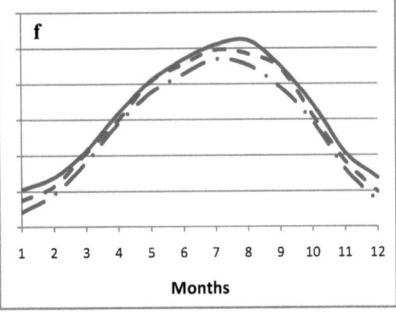

Temperature (°C)

Months

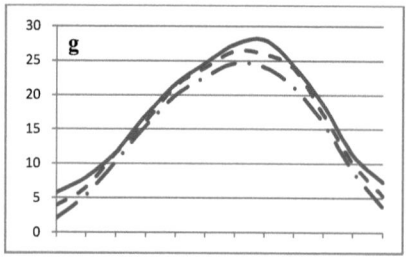
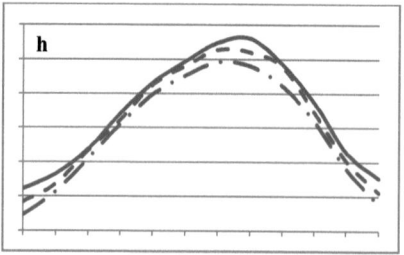
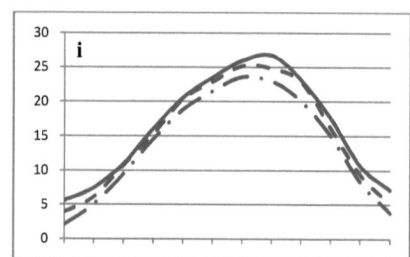
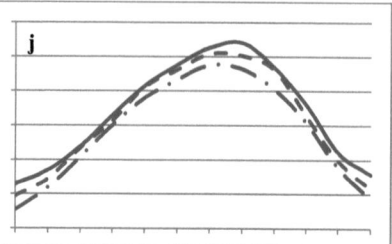
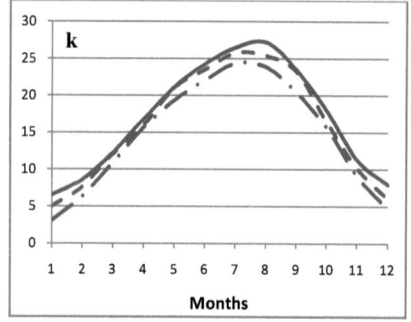
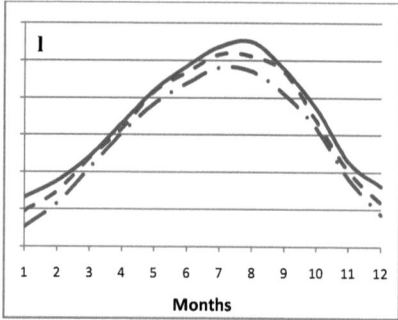

A3. Comparison between the monthly average minimum temperature between the future predicted periods, 2021 – 2050 (—·—), 2051 – 2080 (– – ·), and 2071 – 2100 (——) at at Würzburg (a), Donau-Ries (b), Freising (c), Passau (d), Regensburg (e), Lichtenfels (f), Eichstätt (g), Landshut (h), Neumarkt (i), Weissburg-Gunzenhausen (j), Main-Spessart (k), and Günzburg (l) for REMO model:

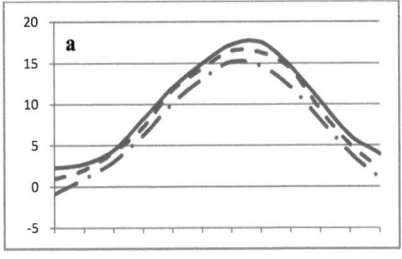
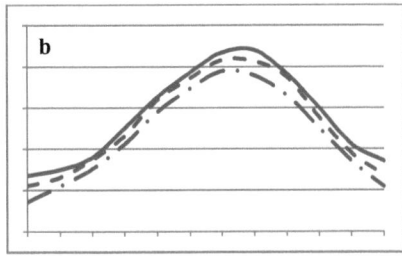
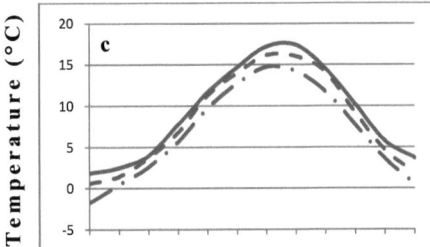
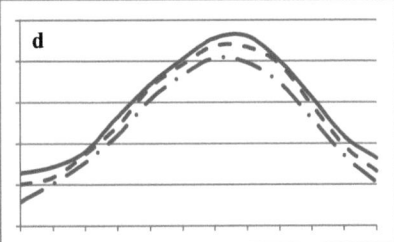
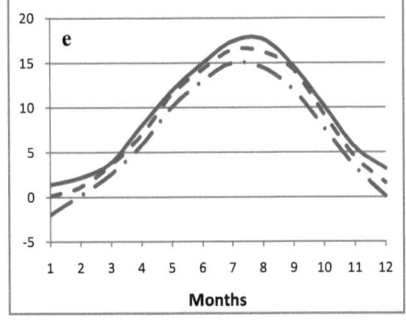
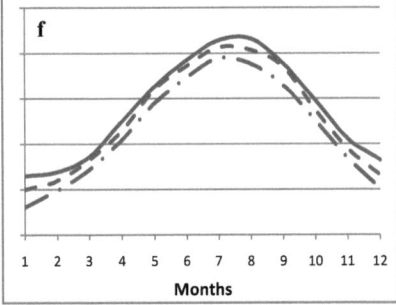

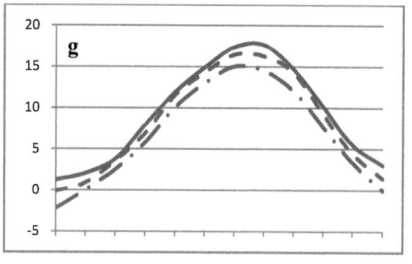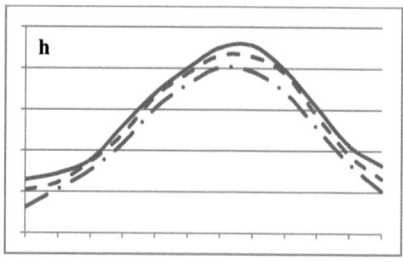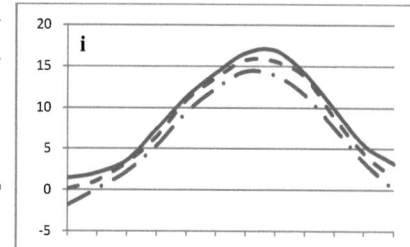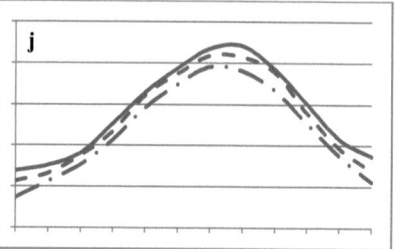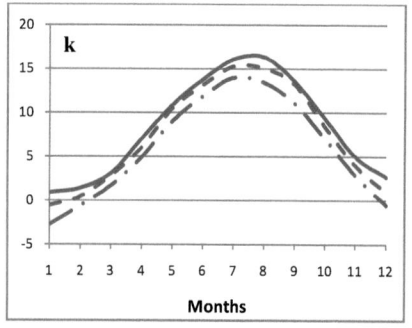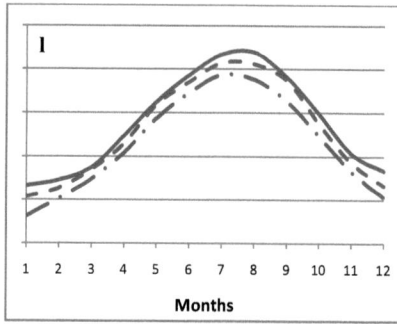

i want morebooks!

Buy your books fast and straightforward online - at one of world's fastest growing online book stores! Environmentally sound due to Print-on-Demand technologies.

Buy your books online at

www.get-morebooks.com

Kaufen Sie Ihre Bücher schnell und unkompliziert online – auf einer der am schnellsten wachsenden Buchhandelsplattformen weltweit! Dank Print-On-Demand umwelt- und ressourcenschonend produziert.

Bücher schneller online kaufen

www.morebooks.de

 VDM Verlagsservicegesellschaft mbH
Heinrich-Böcking-Str. 6-8   Telefon: +49 681 3720 174   info@vdm-vsg.de
D - 66121 Saarbrücken   Telefax: +49 681 3720 1749   www.vdm-vsg.de

Printed by Books on Demand GmbH, Norderstedt / Germany